Mohamed Elzagheid

Organic Chemistry

Also of Interest

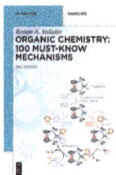

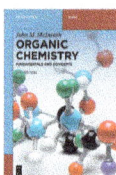

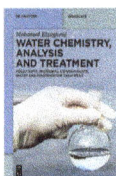

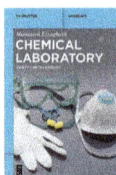

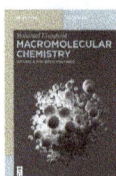

Mohamed Elzagheid

Organic Chemistry

25 Must-Know Classes of Organic Compounds

DE GRUYTER

Author
Prof. Dr. Mohamed Ibrahim Elzagheid
Jubail Industrial College
Chemical Engineering Department
Jubail Industrial City, 31961, KSA
Personal Email: melzagheid@gmail.com
Work Email: elzagheid_m@rcjy.edu.sa

ISBN 978-3-11-138199-2
e-ISBN (PDF) 978-3-11-138275-3
e-ISBN (EPUB) 978-3-11-138330-9

Library of Congress Control Number: 2023951187

Bibliographic information published by the Deutsche Nationalbibliothek
The Deutsche Nationalbibliothek lists this publication in the Deutsche Nationalbibliografie;
detailed bibliographic data are available on the Internet at http://dnb.dnb.de.

© 2024 Walter de Gruyter GmbH, Berlin/Boston
Cover image: jm1366/iStock/Getty Images Plus
Typesetting: Integra Software Services Pvt. Ltd.
Printing and binding: CPI books GmbH, Leck

www.degruyter.com

Preface

Throughout my travels worldwide, studying and working at various universities and on different continents, including Europe, North America, Africa, and Asia, and teaching multiple organic chemistry subjects at the higher education level, I dreamed of compiling my organic chemistry insights into a single short textbook. It is a book that stands out among the hundreds of other organic chemistry books by being succinct, easy to read, and to the point. The book *Organic Chemistry: 25 Must-Know Classes of Organic Compounds* partially meets that dream.

The book begins by discussing several functional groups in organic chemistry, followed by structure and bonding concepts. The book then goes on to introduce the 25 most important classes of organic compounds. These include alkanes, alkenes, dienes, alkynes, cycloalkanes, cycloalkenes, haloalkanes, aromatic hydrocarbons, aryl halides, phenols, alcohols, thiols, ethers, sulfides, heterocyclic compounds containing N, O, and S, aldehydes, ketones, carboxylic acids, carboxylic esters, acid anhydrides, acid halides, amino acids, fatty acids, nitriles, and organic polymers. The book delves further into topics and concepts of stereochemistry and also discusses the most well-known organic reactions in a simple, concise, and straightforward manner while still fulfilling the learner's intellect.

I hope that chemists, academics, and students enjoy reading this book.

Mohamed Ibrahim Elzagheid, Chemistry Professor
Waterloo, Ontario, Canada
2024

https://doi.org/10.1515/9783111382753-202

Acknowledgments

First, I want to thank my entire family for their unwavering support and assistance during my academic journey.

I would also like to thank everyone in the chemical engineering department at Jubail Industrial College. A special thanks goes out to those who joined me in teaching the organic chemistry course at different levels for regular students at Jubail Industrial College and Aramco company trainees in Saudi Arabia as well as my former colleagues at the Faculty of Pharmacy, University of Benghazi in Libya.

Finally, but just as importantly, a huge thank you to the whole publishing team, especially Ute Skambraks, Helene Chavaroche, and Suruthi Manogaran whose help and dedication are immeasurable.

https://doi.org/10.1515/9783111382753-203

The Author

Mohamed Elzagheid is an Associate Professor of Chemistry at the Royal Commission for Jubail and Yanbu as well as a Professor and Consultant for the Libyan Authority for Scientific Research.

During his 30-year career at Turku University in Finland, McGill University, SynPrep Inc. in Montreal, Canada, and Jubail Industrial College in Saudi Arabia, he was directly and indirectly involved in the education of laboratory technicians and chemists and supervised many undergraduate and graduate chemistry students.

He also significantly contributed to numerous short-term and long-term training programs for Saudi companies and taught a variety of university courses at various levels. Organic Chemistry I and II, Polymer Chemistry, Introduction to Macromolecule Chemistry, Biochemistry, Laboratory Techniques, Safety in Chemical Laboratories, Technician Responsibility, and Water and Wastewater Treatment are among the courses offered.

He also chaired the Research, Projects, Publications, and Academic Promotion Team; Academic Promotion Committee; Curriculum Development Committee; Industrial Chemistry Technology Program Advisory and Evaluation Committee; CTAB Steering Accreditation Committee; Industrial Outreach Committee; and Chemical Engineering Department Safety Committee at Jubail Industrial College.

Dr. Elzagheid is the author of six textbooks: *Introductory Organic Chemistry; Thoughts on Organic Chemistry; Macromolecular Chemistry: Natural and Synthetic Polymers; Chemical Laboratory Safety and Techniques; Chemical Technicians: Good Laboratory Practice and Laboratory Information Management Systems;* and *Water Chemistry, Analysis and Treatment: Pollutants, Microbial Contaminants, Water and Wastewater Treatment.*

His work at Turku University in Finland, McGill University in Canada, and JIC in the Kingdom of Saudi Arabia has helped him establish a strong name in chemistry and chemical education, as evidenced by his research papers and publications.

https://doi.org/10.1515/9783111382753-204

Contents

Preface — V

Acknowledgments — VII

The Author — IX

Chapter 1
Introduction — 1
Objectives — 1
1.1   Organic Chemistry — 1
1.2   Functional Groups — 1
1.3   Drawing Chemical Structures — 1
1.4   Branches of Chemistry — 5
1.5   Branches of Organic Chemistry — 6
1.6   Carbon Hybrid Orbitals — 6
1.6.1   The sp^3 Hybrid Orbital — 6
1.6.2   The sp^2 Hybrid Orbital — 7
1.6.3   The sp Hybrid Orbital — 8
1.7   The Nature of Chemical Bonding — 8
1.7.1   Ionic Bonds — 8
1.7.2   Covalent Bonds — 8
1.8   Essential Terms — 10
1.9   Problems — 10

Chapter 2
Aliphatic Hydrocarbons — 14
Objectives — 14
2.1   Introduction — 14
2.2   Naming of Alkanes, Alkenes, Dienes, and Alkynes — 15
2.3   Alkyl Groups — 17
2.4   Cycloalkanes and Cycloalkenes (Alicyclic or Aliphatic Cyclic
            Compounds) — 17
2.5   Haloalkanes (Alkyl Halides) — 18
2.6   Physical Properties of Alkanes, Alkenes, and Alkynes — 20
2.7   Reactions of Alkanes, Alkenes, Dienes, and Alkynes — 20
2.8   Natural Gas and Petroleum — 21
2.9   Polymers Made from Alkenes, Dienes, and Alkene Derivatives — 23
2.10   Essential Terms — 23
2.11   Problems — 25

Chapter 3
Aromatic Hydrocarbons —— 28
Objectives —— 28
3.1   Structure of Benzene —— 28
3.2   Nomenclature (Naming) of Benzene and Benzene Derivatives —— 29
3.3   Polynuclear Aromatic Hydrocarbons —— 31
3.4   Electrophilic Aromatic Substitution (Benzene and Benzene
           Derivatives) —— 31
3.5   Physical Properties of Benzene and Benzene Derivatives —— 32
3.6   Preparation and Reactions of Benzene Derivatives —— 33
3.6.1   Halobenzenes (Aryl Halides) —— 34
3.6.2   Hydroxybenzenes (Phenols) —— 34
3.7   Industrial Applications of Benzene and Benzene Derivatives —— 36
3.8   Essential Terms —— 37
3.9   Problems —— 37

Chapter 4
Amines, Alcohols, Thiols, Ethers, Sulfides, and Heterocyclic Compounds
Containing Nitrogen, Oxygen, and Sulfur —— 40
Objectives —— 40
4.1   Amines —— 40
4.1.1   Physical Properties of Amines —— 40
4.1.2   Preparation of Amines —— 41
4.1.3   Industrial Sources —— 42
4.1.4   Reactions of Amines —— 42
4.2   Alcohols —— 43
4.2.1   Structure and Nomenclature of Alcohols —— 43
4.2.2   Physical Properties of Alcohols —— 43
4.2.3   Preparation of Alcohols —— 44
4.2.4   Reactions of Alcohols —— 45
4.2.5   Polyhydroxy Alcohols —— 45
4.3   Thiols (Mercaptans) —— 45
4.3.1   Structure and Nomenclature of Thiols —— 45
4.3.2   Physical Properties of Thiols —— 46
4.3.3   Preparation and Reactions of Thiols —— 46
4.4   Ethers —— 48
4.4.1   Structure and Nomenclature of Ethers —— 48
4.4.2   Physical Properties of Ethers —— 48
4.4.3   Preparation of Ethers —— 50
4.4.4   Reactions of Ethers —— 50
4.5   Sulfides —— 51

4.5.1   Structure and Nomenclature of Sulfides —— **51**
4.5.2   Preparation and Reactions of Sulfides —— **51**
4.6   Heterocyclic Compounds Containing Nitrogen, Oxygen,
              and Sulfur —— **51**
4.6.1   Nomenclature of Heterocyclic Compounds —— **54**
4.6.2   Reactions of Heterocyclic Compounds —— **57**
4.7   Essential Terms —— **58**
4.8   Problems —— **59**

Chapter 5
Aldehydes and Ketones —— 62
Objectives —— **62**
5.1   Structure and Nomenclature of Aldehydes and Ketones —— **62**
5.2   Physical Properties of Aldehydes and Ketones —— **62**
5.3   Preparation of Aldehydes and Ketones —— **63**
5.4   Reactions of Aldehydes and Ketones —— **65**
5.5   Essential Terms —— **68**
5.6   Problems —— **69**

Chapter 6
Carboxylic Acids and Their Derivatives —— 71
Objectives —— **71**
6.1   Carboxylic Acids —— **71**
6.1.1   Structure and Nomenclature of Carboxylic Acids —— **71**
6.1.2   Physical Properties of Carboxylic Acids —— **73**
6.1.3   Long-Chain Carboxylic Acids (Fatty Acids) —— **74**
6.1.4   Amino Carboxylic Acids (Amino Acids) —— **75**
6.1.5   Reactions of Carboxylic Acids —— **76**
6.2   Dicarboxylic Acids —— **76**
6.3   Carboxylic Acid Derivatives —— **78**
6.3.1   Structure and Nomenclature of Carboxylic Acids Derivatives —— **78**
6.3.2   Physical Properties of Carboxylic Acids Derivatives —— **79**
6.3.3   Reactions of Carboxylic Acids Derivatives —— **80**
6.4   Essential Terms —— **82**
6.5   Problems —— **83**

Chapter 7
Organic Polymers —— 85
Objectives —— **85**
7.1   Notation and Nomenclature of Organic Polymers —— **85**
7.2   Polymers Architecture —— **85**

7.3   Polymers Morphology —— **86**
7.4   Polymerization Reactions —— **87**
7.5   Polymer Stereochemistry (Tacticity) —— **89**
7.6   Commodity Polymers —— **89**
7.7   Copolymers —— **90**
7.8   Essential Terms —— **91**
7.9   Problems —— **92**

Chapter 8
Stereochemistry Topics and Concepts —— 94
Objectives —— **94**
8.1   Stereochemistry and Isomers —— **94**
8.2   The Homologous Series —— **95**
8.2.1   Definition —— **95**
8.2.2   Characteristics of Homologous Series —— **96**
8.3   Molecular Chirality —— **96**
8.4   Stereocenter and Chiral Center —— **98**
8.5   The E-Z Notational System —— **98**
8.6   The R-S Notational System —— **99**
8.7   Racemic Mixture —— **99**
8.8   Meso Compounds —— **102**
8.9   D-L Isomers —— **103**
8.10   Epimers and Epimerization —— **103**
8.11   Syn-Anti Isomers —— **104**
8.12   Cis-Trans Isomers —— **105**
8.13   Exo-Endo Isomerism —— **105**
8.14   Tautomerism —— **105**
8.15   Essential Terms —— **106**
8.16   Problems —— **107**

Chapter 9
The Most Common Organic Reactions —— 111
Objectives —— **111**
9.1   Carbonyl Group Involved Organic Reactions —— **111**
9.2   Organic Reactions Involve Carbon-Carbon Double and
           Triple Bonds —— **111**
9.3   Organic Reactions Involve Alkyl Halides and Aryl Halides —— **117**
9.4   Organic Reactions Involve Alcohols —— **120**
9.5   Essential Terms —— **124**
9.6   Problems —— **124**

Solutions to Problems —— 126

Abbreviations —— 159

Resources and Further Readings —— 161

Appendix A: List of Selected Important Organic Compounds —— 165

Appendix B: List of Selected Organic Reagents —— 173

Index —— 177

Chapter 1
Introduction

Objectives

After studying this chapter, learners will be able to:
- Write the functional group structures of organic compounds.
- Draw chemical structures of certain organic compounds by Kekulé, condensed, and skeletal formulas.
- Know chemistry branches.
- Know organic chemistry subbranches.
- Know the carbon atom hybrid orbitals.
- Have an idea about chemical bonds.

1.1 Organic Chemistry

Organic chemistry is the chemistry of carbon compounds because carbon atoms are there in all organic compounds. They form four covalent bonds in organic structures and can exist in long chains by connecting their atoms one to another. Carbon can form a large diverse number of compounds, from simple structures like methane and ethane to complex structures like ribonucleic acids, deoxyribonucleic acids, and polypeptides. Figure 1.1 shows other elements commonly found in organic compounds and bond to carbon atoms.

1.2 Functional Groups

Functional groups are specific groups of atoms responsible for the characteristic chemical reactions of the compounds. Groups (*only the alkyl or aliphatic ones are represented here but R can be replaced with Ar to get the aryl or aromatic ones*) that are commonly found in organic chemistry are shown in Figure 1.2A.

A functional group can exist as a single group in a compound or in combination with other groups in the structure. This is seen in the chemical organic structures shown in Figure 1.2B.

1.3 Drawing Chemical Structures

Three different formulas are used to draw organic molecules. These include Kekulé, Condensed, and Skeletal (line) formulas. In the Kekulé formula atoms of the molecule

https://doi.org/10.1515/9783111382753-001

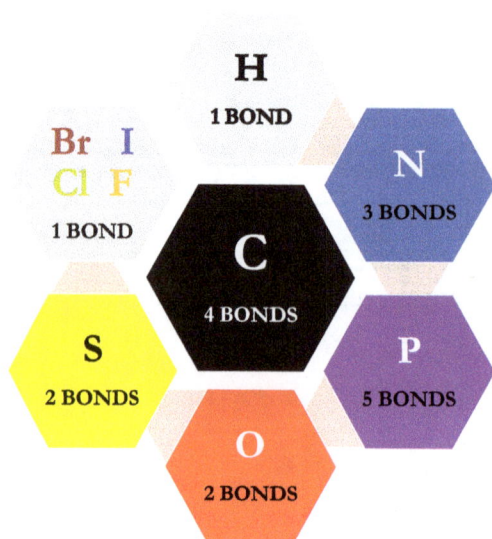

Figure 1.1: Elements that are commonly bond to carbon and found in organic compounds.

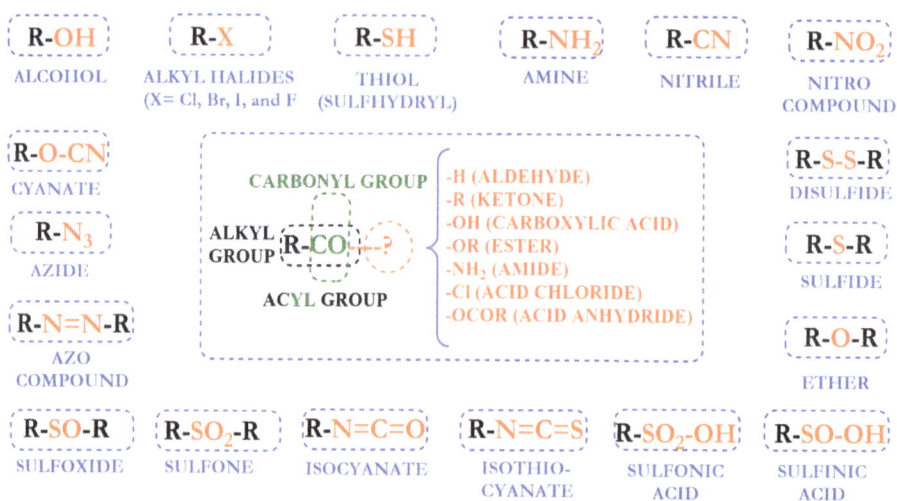

Figure 1.2A: Functional groups in organic chemistry.

are shown in the order they are bonded, and all bonds are presented in the structure. The condensed formula shows the connectivity and the order of atoms in a single line. The same groups are surrounded by parenthesis and a subscript number. The number reflects how many groups are attached to the same atom.

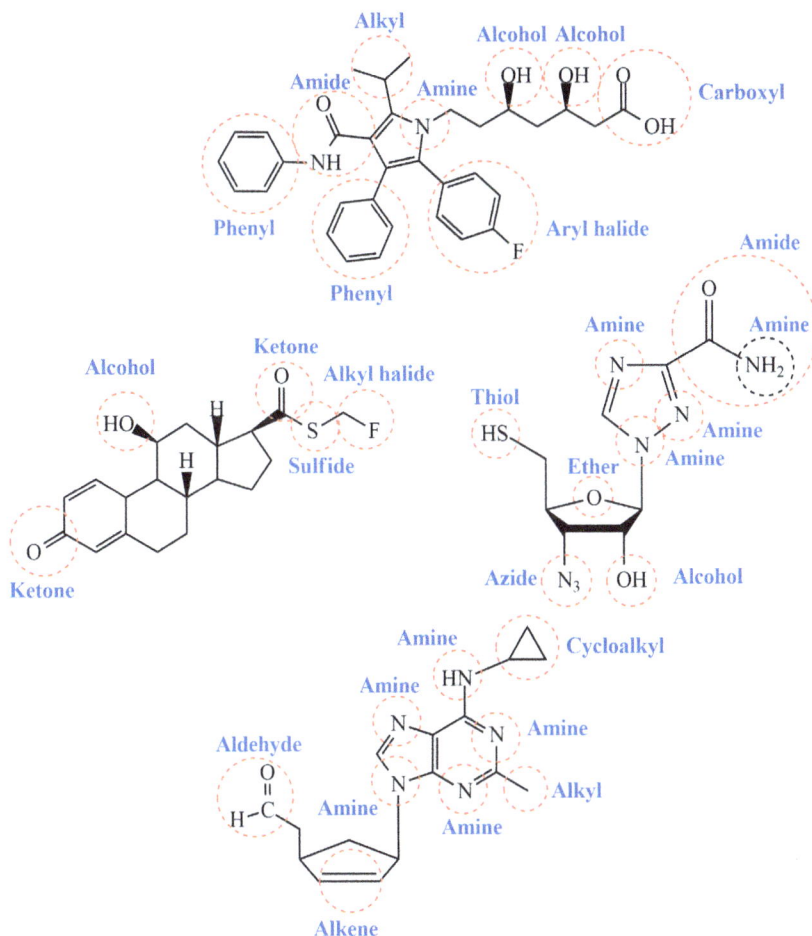

Figure 1.2B: Multiple functional groups in different organic compounds.

In the line-angle formula, carbon atoms are replaced with lines and hydrogen atoms are neglected. Examples of selected organic families represented by Kekulé, condensed, and skeletal formulas are shown in Table 1.1.

Table 1.1: Kekulé, condensed, and skeletal formulas of selected organic compounds.

Organic family (example)	Kekulé formula	Condensed formula	Line (skeletal) formula
Alkanes (propane)		$CH_3CH_2CH_3$	
Alkenes (propene)		$CH_2=CHCH_3$	
Alkynes (propyne)		$CH\equiv CCH_3$	
Arenes (benzene)		C_6H_6	
Alcohols (propanol)		$CH_3CH_2CH_2OH$	
Alkyl halides (propyl chloride)		$CH_3CH_2CH_2Cl$	
Aryl halide (phenyl chloride)		C_6H_5Cl	
Aldehyde (propanal)		CH_3CH_2CHO	

Table 1.1 (continued)

Organic family (example)	Kekulé formula	Condensed formula	Line (skeletal) formula
Ketone (propanone)		CH₃COCH₃	
Carboxylic acid (propanoic acid)		CH₃CH₂COOH	

1.4 Branches of Chemistry

Chemistry is divided into several subfields or specializations. Analytical chemistry, physical chemistry, biochemistry, organic chemistry, and inorganic chemistry are the five major fields of chemistry (Figure 1.3). In contrast, organic chemistry is sometimes thought to contain biochemistry as a subfield. Physics, biology, and chemistry all intertwine or overlap. There is also some crossover with engineering. Each major discipline has multiple subcategories.

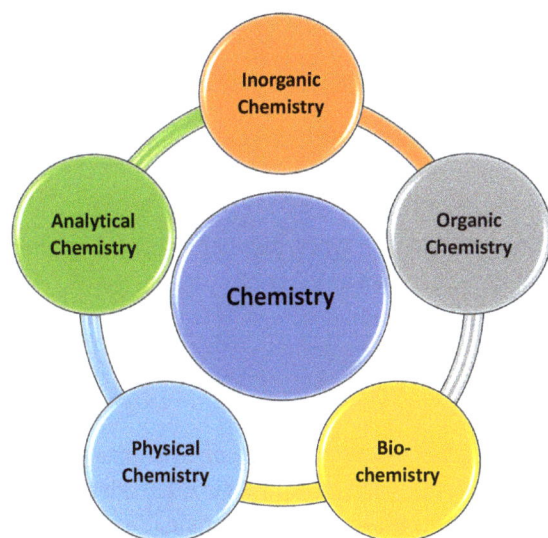

Figure 1.3: Branches of chemistry.

1.5 Branches of Organic Chemistry

The knowledge of organic chemistry is crucial to many scientists. In fact, all areas of study dealing with petroleum, natural gas, medicines, plants, animals, and even microorganisms are dependent on the basics of organic chemistry. Due to that many branches of organic chemistry already exist and new ones will follow. Selected branches of organic chemistry are briefly presented in Figure 1.4.

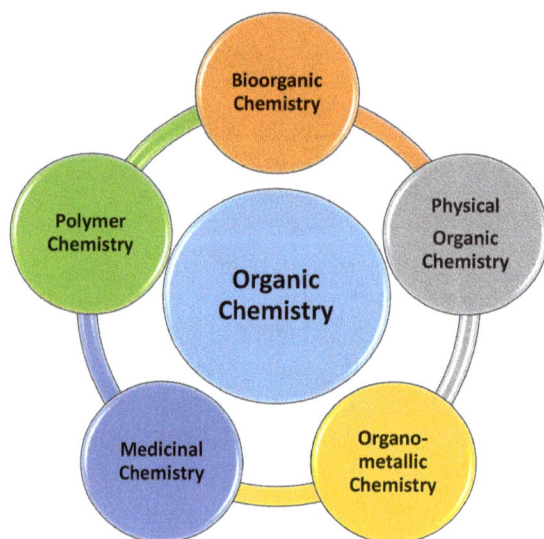

Figure 1.4: Branches of organic chemistry.

1.6 Carbon Hybrid Orbitals

The pure carbon atomic orbitals, their hybridization types, number of hybrid orbitals produced for each of them, and examples are shown in Table 1.2.

1.6.1 The sp³ Hybrid Orbital

In the sp^3 hybrids, an s orbital and three p orbitals can combine or hybridize to form four equivalent atomic orbitals with tetrahedral orientation (Figure 1.5). Please note that the small lobes of the sp^3 orbitals are not shown. A good example of sp^3 carbons can be seen in the ethane carbons.

Table 1.2: Different types of hybrid orbitals.

Pure atomic orbitals of the central atom	Hybridization of the central atom	Number of hybrid orbitals	Examples
s, p, p, p	sp^3	4	
s, p, p	sp^2	3	
s, p	sp	2	

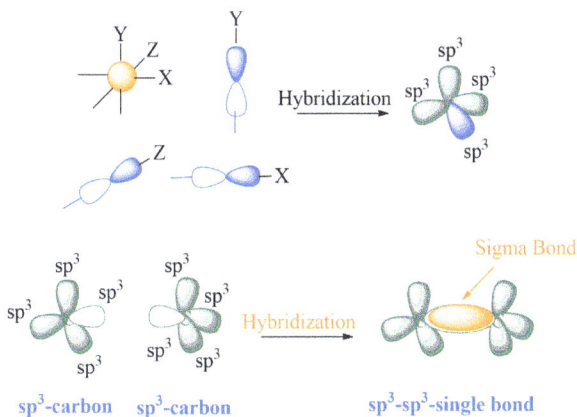

Figure 1.5: Sp^3 hybrid orbital and formation of sp^3-sp^3 carbon-carbon single bond.

1.6.2 The sp^2 Hybrid Orbital

In the sp^2 hybrids, there are three equivalent sp^2 hybrid orbitals lying in a plane at angles of 120° to each other and a single unhybridized p orbital perpendicular to the sp^2 plane (Figure 1.6). The sp^2 carbons can be seen in ethylene carbon-carbon double bond.

Pi-Bond

sp² p sp² sp² sp² Hybridization

sp² p sp² p

Sigma Bond

sp²-carbon   **sp²-carbon**   **sp²sp²-double bond**

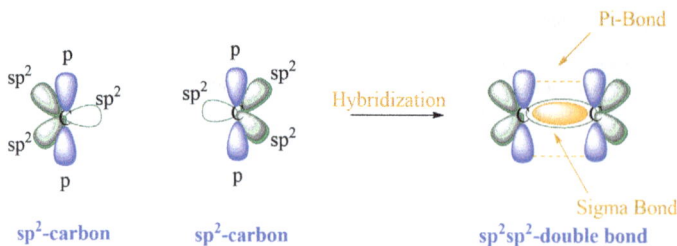

Figure 1.6: Sp² hybrid orbital and formation of sp²-sp² carbon-carbon double bond.

1.6.3 The sp Hybrid Orbital

The sp hybrid involves the combining of a carbon 2s orbital with only a single p orbital. Two sp orbitals result and two p orbitals remain unchanged (Figure 1.7). The sp carbons can be seen in ethyne carbon-carbon triple bond.

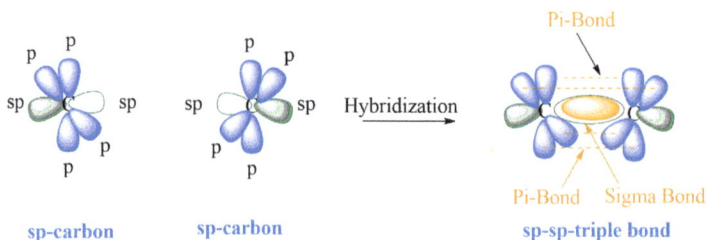

Pi-Bond

p p sp sp sp p p sp Hybridization

p p p

Pi-Bond   Sigma Bond

sp-carbon   **sp-carbon**   **sp-sp-triple bond**

Figure 1.7: Sp hybrid orbital and formation of sp-sp carbon-carbon triple bond.

1.7 The Nature of Chemical Bonding

1.7.1 Ionic Bonds

Electrovalent (ionic) bonds are formed by an electrical attraction between positively charged cations and negatively charged anions (Figure 1.8).

1.7.2 Covalent Bonds

Covalent bonds are formed by sharing electrons between atoms. The number of covalent bonds an atom forms depends on how many electrons it can have in its valence examples are shown in Figure 1.9.

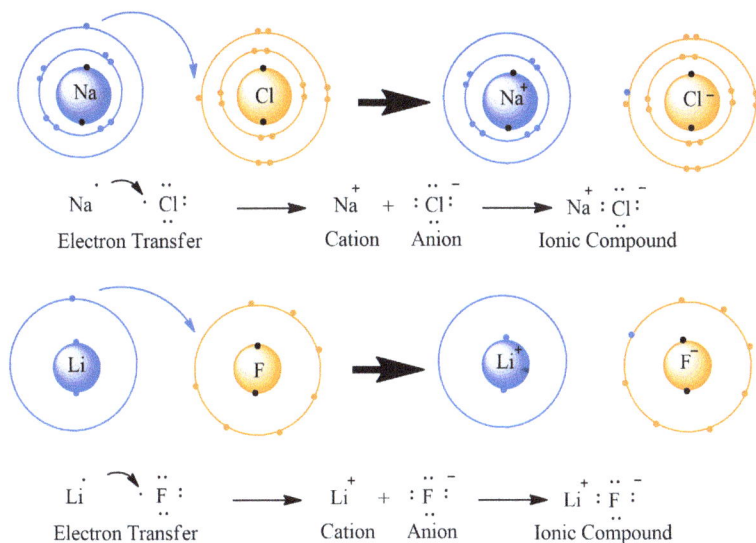

Figure 1.8: Examples of ionic bonds in sodium chloride and lithium fluoride.

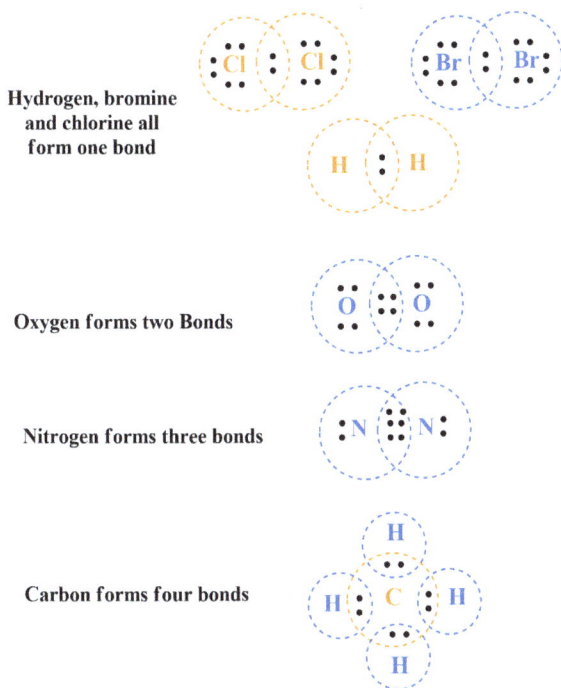

Figure 1.9: Covalent bonds examples.

1.8 Essential Terms

Atomic orbitals Computer-generated representations that describe the location, shape, and behavior of electrons in atoms.

Condensed structure A way of writing organic chemical structures in a line of text of all atoms, but ignores the vertical bonds and most or all the horizontal single bonds.

Covalent bond The bond formed as a result of sharing electrons between atoms.

Electrovalent bond The bond formed by an electrical attraction between positively charged cations and negatively charged anions.

Functional groups The families of organic compounds have the same structural features and similar chemical reactivity.

Hybridization The mixing of pure atomic orbitals to form a set of new hybrid orbitals.

Kekulé structure A structure in which bonded electron pairs in covalent bonds are drawn as lines.

Organic chemistry The chemistry of carbon compounds due to the existence of carbon in all organic compounds

Skeletal structure A representation of molecular structure in which covalent bonds are drawn as lines and carbon atoms are shown as angles.

1.9 Problems

1.9.1   Identify at least two functional groups in the following organic compound:

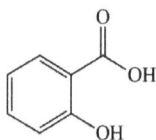

1.9.2   Convert the following skeletal structures into Kekulé structures:

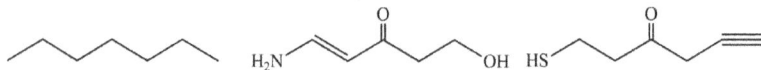

1.9.3   Draw the structures of the compound with the following molecular formulas:
   i. $C_4H_{10}O$
   ii. C_4H_8O
   iii. $C_4H_8O_2$

1.9.4   How many bonds can the following atoms form?
   i.   Carbon
   ii.   Sulfur

iii. Nitrogen

iv. Oxygen

1.9.5   If the condensed structure for 2-butanone is $CH_3CH_2COCH_3$, show how the Kekulé structure looks like.

1.9.6   Draw the skeletal formulas for all possible compounds with the molecular formula C_4H_8O.

1.9.7   Draw the functional group structure for the following families of organic compounds:

Family name	Functional group structure
Isocyanates	
Nitriles	
Sulfoxides	

1.9.8   Draw the Kekulé structures of the following molecules:

i. $HO (CH_2)_4NH_2$

ii. $(CH_3)_3CCCl_3$

iii. $(CH_3)_3CCH_2CHO$

1.9.9   Complete these Kekulé formulas by adding enough hydrogens to complete the tetravalence of each carbon (carbon forms four bonds).

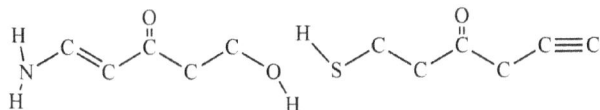

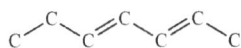

1.9.10 Write the family name for the following functional group structures:

Functional group structure	Family name
R-COOH	
R-N=N-R	
$RCONH_2$	

1.9.11 How many carbon atoms are present in the following molecule?

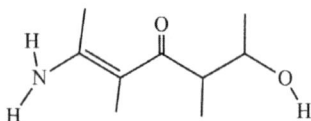

1.9.12 How many hydrogen atoms are there in the following structure?

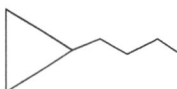

1.9.13 Which of the following compounds can form covalent bonds?
   i. CH_4
   ii. NaCl
   iii. NH_3

1.9.14 Which of the following molecule bonding patterns is not correct in organic chemistry?

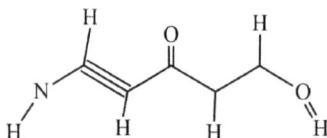

1.9.15 Which one of the following condensed formulas represents the same compound as in the skeletal formula shown below?

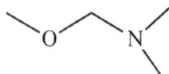

   i. $CH_3OCH_2N(CH_3)_2$
   ii. $CH_3OCH_2N(CH_2CH_3)_2$

1.9.16 How many carbon-carbon σ bonds are present in the following structure?

1.9.17 How many π bonds are present in the molecule shown?

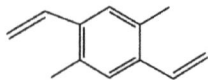

1.9.18 Indicate the hybridization of each carbon in the following structures:

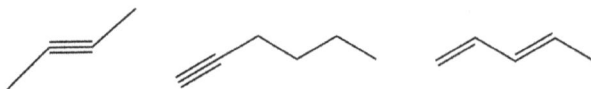

1.9.19 Draw the chemical structure of propanone.
1.9.20 Does oxygen molecule show igma bonding or pi bonding or both?

1.9.21 If the condensed structure for 2-hexanone is $CH_3CH_2CH_2CH_2COCH_3$, show how the skeletal structure looks like.

1.9.22 Draw the condensed formulas for all possible compounds with the molecular formula $C_6H_{12}O$.

1.9.23 Complete the missing information in the following table:

Pure atomic orbitals of the central atom	Hybridization of the central atom
s, p, p	
s, p	

1.9.24 How many sp^2 and sp-hybridized carbon atoms are present in the following structure?

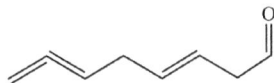

1.9.25 What kind and how many pure atomic orbitals are involved in the formation of sp2 hybrid orbital?

1.9.26 Convert the following skeletal structure into condensed and Kekulé structures.

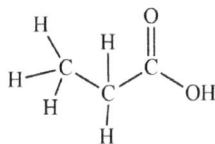

1.9.27 Which of these possible structural formulas for $C_3H_6O_2$ is correct?

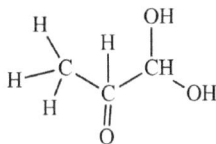

1.9.28 Vinegar contains acetic acid as a major organic constituent. Convert the following Kekulé formula of acetic acid into a skeletal structure.

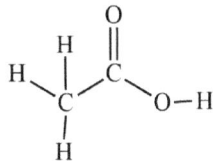

1.9.29 The following formulas are incorrect. Find out which is wrong with each one.
  i. CCl_3
  ii. H_3S
  iii. N_3H_4

1.9.30 Draw the skeletal formulas for two compounds having the molecular formula C_2H_4O.

****

Chapter 2
Aliphatic Hydrocarbons

Objectives

After studying this chapter, learners will be able to:
- Write the common and IUPAC names of alkanes, haloalkanes, alkenes, dienes, and alkynes.
- Draw the chemical structures of aliphatic hydrocarbons and haloalkanes.
- Describe the important methods of preparation and reactions of these classes of compounds.
- Correlate their physical properties.
- Discuss natural gas and petroleum and list examples of polymers produced from alkenes and alkene derivatives.

2.1 Introduction

Hydrocarbons are a class of organic compounds that have only hydrogen and carbon in their structures. They can be classified as "saturated" where compounds have only carbon-carbon single bonds or "unsaturated" where compounds have one carbon-carbon pi-bond in addition to single bonds. Alkanes are examples of saturated hydrocarbons and alkenes, alkynes, and arenes represent unsaturated hydrocarbons (Figure 2.1).

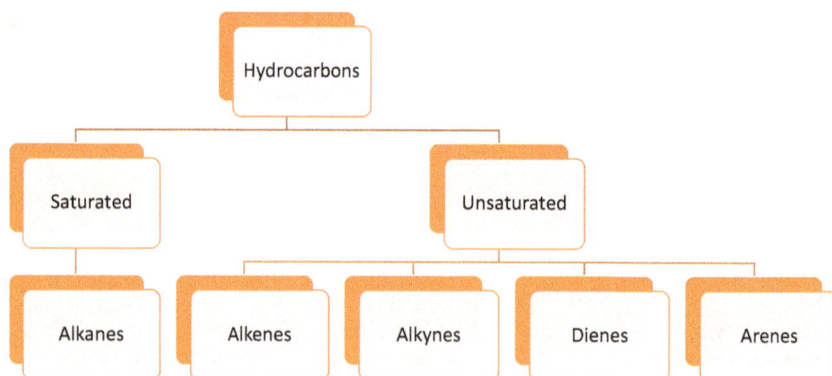

Figure 2.1: Saturated and unsaturated hydrocarbons.

All above-mentioned hydrocarbons are aliphatic and those contain benzene rings, or their derivatives are known as aromatic hydrocarbons (Figure 2.2).

https://doi.org/10.1515/9783111382753-002

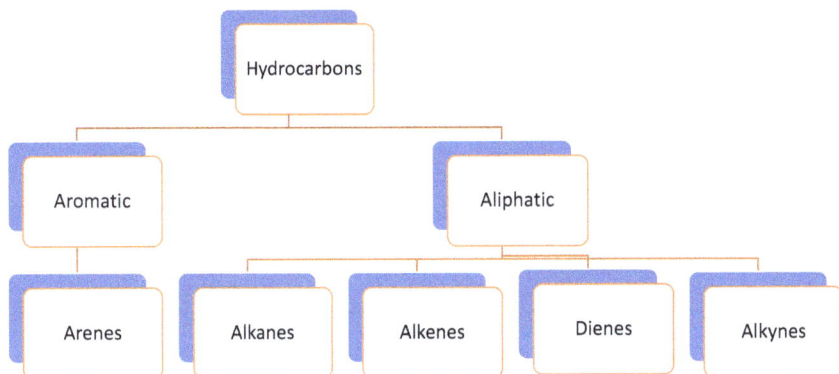

Figure 2.2: Aliphatic and aromatic hydrocarbons.

2.2 Naming of Alkanes, Alkenes, Dienes, and Alkynes

Alkanes are saturated (they have only C–C and C–H single bonds) hydrocarbons (contain only carbon and hydrogen). Their general formula is C_nH_{2n+2}, where n is an integer (number). The first three simple alkenes are named (according to the common naming system) as ethylene, propylene, and isobutylene. Alkenes are named using the IUPAC naming system by replacing the -ane ending of the corresponding alkane with -ene ending. The longest continuous chain should contain the C=C double bond and be considered the parent or the base name. The C=C is given the lowest possible number, regardless of what other alkyl substituents are present. But if an OH group is present, it outranks the double bond. Numbers are assigned to the substituents and also to the first carbon in the C=C double bond. Alkynes have a general chemical formula of C_nH_{2n-2}, and they contain carbon-carbon triple bonds in addition to carbon-carbon single bonds in their structure. Unlike alkenes, alkynes can either have a terminal or internal triple bond in their structures and this, in turn, has an effect on their physical properties. The suffix -yne is used to denote an alkyne and the position of the triple bond is indicated by giving the number of the first alkyne carbon in the chain.

Numbering the main chain begins at the end nearer the triple bond so that the triple bond receives as low number as possible. If more than one triple bond exists the alkynes are called diynes, triynes, and so forth.

If both double and triple bonds are existing the alkynes are called enynes. The numbering of enyne chains starts from the end closed to the first multiple bonds. When there is a choice in numbering, double bond receives lower numbers than triple bonds.

If compounds contain hydroxyl group and carbon-carbon triple bonds the OH group is given the lowest number. Dienes are another type of unsaturated hydrocarbons that have different arrangements of double bonds. In the conjugated dienes, the two double bonds are separated by just one single bond. The isolated dienes have the two double bonds separated by two or more single bonds. On the other hand, in the cumulated dienes, the two double bonds are successive with no intervening single bonds. Selected examples of straight-chain and branched alkanes, alkenes, dienes, and alkynes are presented in Figure 2.3, and alkanes, alkenes, and alkynes containing 1–20 carbon atoms are listed in Tables 2.1.

Table 2.1: Alkanes, alkene, and alkynes 1–10 carbons.

Number of carbons	Names of alkane/alkene/alkyne	Formula $(CnH_{2n+2})/(C_nH_{2n})/(C_nH_{2n-2})$
1	Methane	CH_4
2	Ethane/ethene/ethyne	$C_2H_6/C_2H_4/C_2H_2$
3	Propane/propene/propyne	$C_3H_8/C_3H_6/C_3H_4$
4	Butane/butene/butyne	$C_4H_{10}/C_4H_8/C_4H_6$
5	Pentane/pentene/pentyne	$C_5H_{12}/C_5H_{10}/C_5H_8$
6	Hexane/hexene/hexyne	$C_6H_{14}/C_6H_{12}/C_6H_{10}$
7	Heptane/heptene/heptyne	$C_7H_{16}/C_7H_{14}/C_7H_{12}$
8	Octane/octene/octyne	$C_8H_{18}/C_8H_{16}/C_8H_{14}$
9	Nonane/nonene/nonyne	$C_9H_{20}/C_9H_{18}/C_9H_{16}$
10	Decane/decene/decyne	$C_{10}H_{22}/C_{10}H_{20}/C_{10}H_{18}$

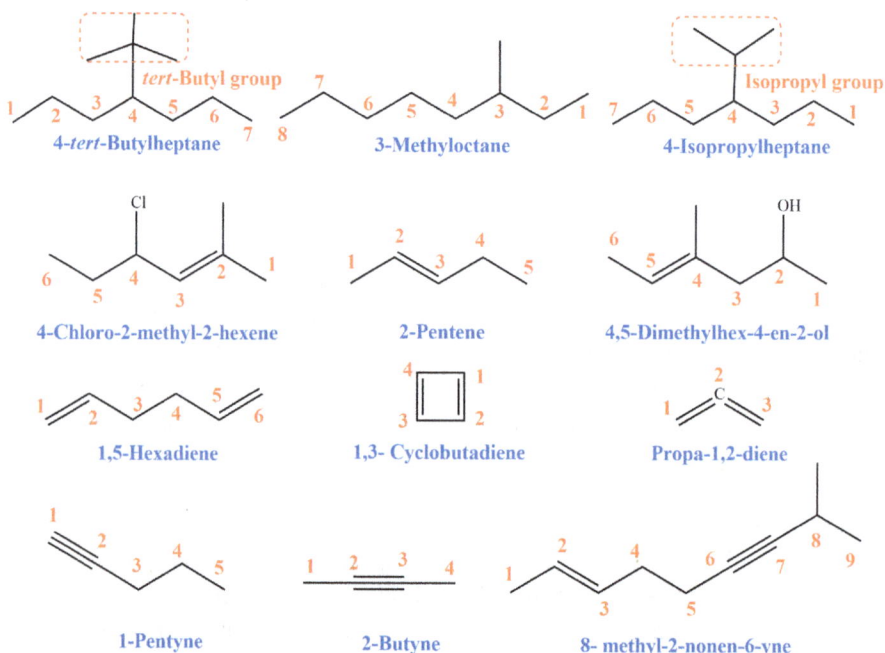

Figure 2.3: Selected examples of straight-chain and branched alkanes, alkenes, dienes, and alkynes.

2.3 Alkyl Groups

An alky group is an alkane with one less hydrogen atom (alkane-H=alkyl). It is named by replacing -ane in an alkane with -yl (Table 2.2). The symbol (R) is used to represent a generalized alkyl group. The (R) group can be methyl, ethyl, propyl, or any other alkyl group.

Table 2.2: Alkyl groups.

Alkane	Formula	Alkyl (alkane-H)	Formula
Methane	CH_4	Methyl	CH_3
Ethane	CH_3CH_3 or (C_2H_6)	Ethyl	C_2H_5
Propane	$CH_3CH_2CH_3$ or (C_3H_8)	Propyl	C_3H_7
Butane	C_4H_{10}	Butyl	C_4H_9
Pentane	C_5H_{12}	Pentyl	C_5H_{11}
Hexane	C_6H_{14}	Hexyl	C_6H_{13}
Heptane	C_7H_{16}	Heptyl	C_7H_{15}

2.4 Cycloalkanes and Cycloalkenes (Alicyclic or Aliphatic Cyclic Compounds)

Cycloalkanes are represented by polygons in skeletal drawings and have the general formula C_nH_{2n}. Alkyl monosubstituted cycloalkanes are named by counting the number of carbons in the ring and the number in the largest substituent.

If the ring has more carbons than the chain (the substituent), the ring takes the parent name but, if the chain has more carbons or the same carbons, the ring will be considered the substituent. For alkyl di-, tri-, tetra-substituted cycloalkanes the numbering of the substituents on the ring should be done so as to arrive at the lowest sum. The same procedure used for naming cycloalkanes can be followed for naming cycloalkenes considering the presence of the double bonds in the structure cycloalkenes. The general formula of cycloalkenes is $C_nH_{2(n-m)}$, where the letter m represents the number of double bonds. The numbering begins with the double bond similar to alkenes and continues around the ring. Selected examples of cycloalkanes and cycloalkenes are presented in Figure 2.4.

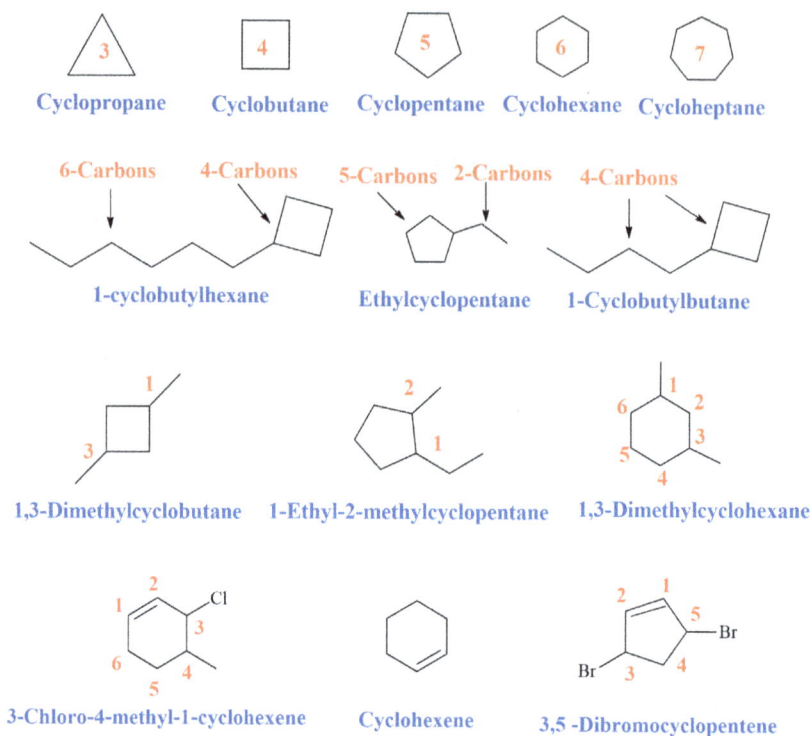

Figure 2.4: Selected examples of cycloalkanes and cycloalkenes.

2.5 Haloalkanes (Alkyl Halides)

Haloalkanes have the general formula R–X, in which R is an alkyl or a substituted alkyl group. The halogen atom (X) is the functional group and reactions of haloalkanes take place at the halogen atom. Carbon atoms in haloalkanes are classified as primary (1°), secondary (2°), or tertiary (3°), according to the number of other carbon atoms (or alkyl groups) attached to them (Figure 2.5).

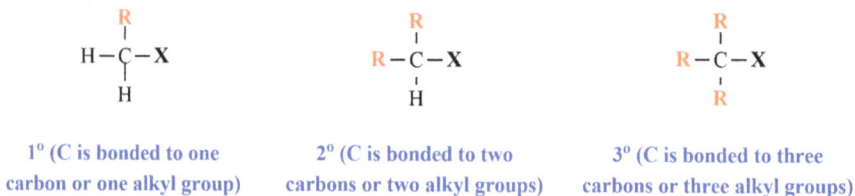

1° (C is bonded to one carbon or one alkyl group)

2° (C is bonded to two carbons or two alkyl groups)

3° (C is bonded to three carbons or three alkyl groups)

Figure 2.5: Classes of haloalkanes.

Haloalkanes can be named by a common naming system (for simpler halides) or an IUPAC naming system. Examples of both naming systems are listed below in Figure 2.6.

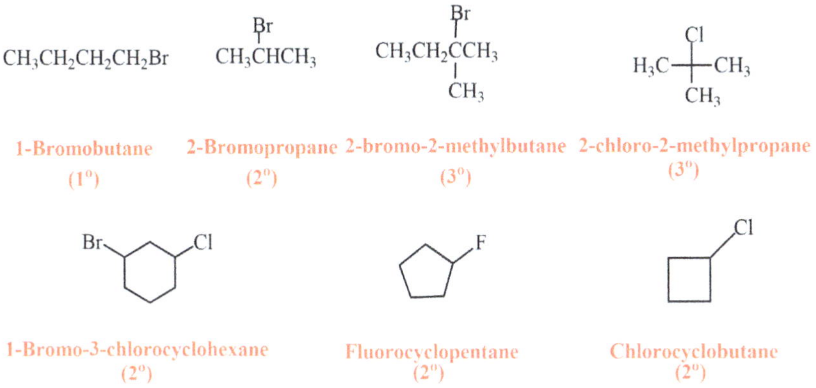

CH$_3$CH$_2$CH$_2$CH$_2$Br

$\overset{\text{Br}}{\underset{|}{\text{CH}_3\text{CHCH}_3}}$

$\overset{\text{Br}}{\underset{\underset{\text{CH}_3}{|}}{\overset{|}{\text{CH}_3\text{CH}_2\text{CCH}_3}}}$

$\text{H}_3\text{C}\overset{\overset{\text{Cl}}{|}}{\underset{\underset{\text{CH}_3}{|}}{\rule{0pt}{0pt}}}\text{CH}_3$

1-Bromobutane
(1°)

2-Bromopropane
(2°)

2-bromo-2-methylbutane
(3°)

2-chloro-2-methylpropane
(3°)

1-Bromo-3-chlorocyclohexane
(2°)

Fluorocyclopentane
(2°)

Chlorocyclobutane
(2°)

Figure 2.6: Examples of alkyl halides.

Haloalkanes can be prepared from alcohols by reacting them with thionyl chloride, phosphorous pentachloride, and HXs (Figure 2.7).

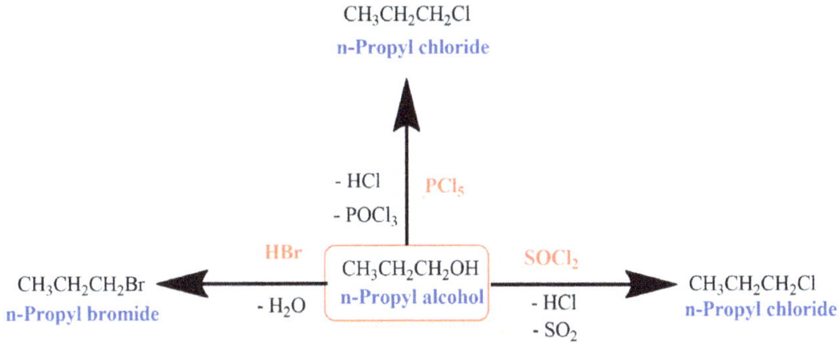

CH$_3$CH$_2$CH$_2$Cl
n-Propyl chloride

- HCl   PCl$_5$
- POCl$_3$

HBr

CH$_3$CH$_2$CH$_2$Br
n-Propyl bromide

- H$_2$O

CH$_3$CH$_2$CH$_2$OH
n-Propyl alcohol

SOCl$_2$

- HCl
- SO$_2$

CH$_3$CH$_2$CH$_2$Cl
n-Propyl chloride

Figure 2.7: Preparation of haloalkanes from alcohol by different reagents.

They are also prepared by direct halogenation of hydrocarbons with chlorine or bromine in the presence of heat or light. Another method of their preparation is by the Markovnikov addition of hydrogen halides to alkenes to give monohalogenated alkenes or by the addition of halogens to alkenes to get vicinal alkyl dihalides as shown in Figure 2.8.

2.6 Physical Properties of Alkanes, Alkenes, and Alkynes

Alkane, alkenes, and alkynes are quite soluble in non-polar solvents such as ether and chloroform but insoluble in water and the boiling points increase as the number of carbons increases. They are also less dense than water. Terminal alkynes (e.g., 1-butyne) have a lower boiling point than internal alkynes (e.g., 2-butyne).

2.7 Reactions of Alkanes, Alkenes, Dienes, and Alkynes

Alkanes undergo three reaction types: halogenation, combustion, and pyrolysis. Alkenes and alkynes are generally more reactive than alkanes due to their electron density in the pi bonds and different addition reactions can take place across their double and triple bonds. Dienes with 1,3-double bonds arrangement can participate in the Diels-Alder reaction to give cyclohexenes. In the presence of the hydrogenations catalysts such as platinum, palladium, and nickel, hydrogen can be added to the triple or the double bond to form an alkene from an alkyne or an alkane from an alkene. Alkenes and alkynes can also be halogenated with the halogen adding across the double or triple bond in a similar fashion to hydrogenation. The halogenation of an alkene results in a dihalogenated alkane product, while the halogenation of an alkyne can

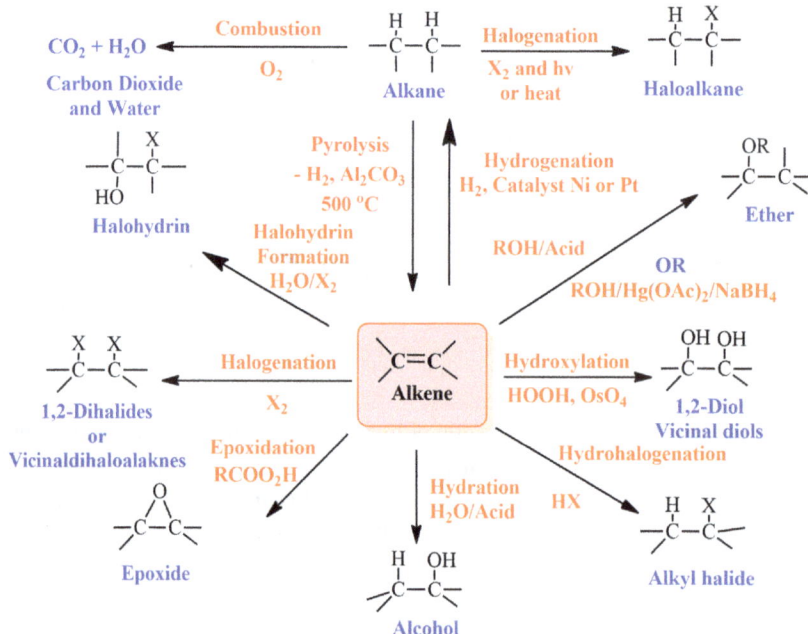

Figure 2.8: Reactions of alkanes and alkenes.

produce a tetrahalogenated alkane. They can also react with hydrogen halides like HCl and HBr. This hydrohalogenation reaction gives the corresponding vinyl halides or alkyl dihalides, depending on the number of HX equivalents added. Water can also be added across triple bonds in alkynes to yield aldehydes and ketones.

Hydration of alkenes via oxymercuration or by acid catalyst produces alcohols and halogenation in the presence of water gives halohydrins. All above-mentioned reactions are summarized in Figures 2.8 and 2.9.

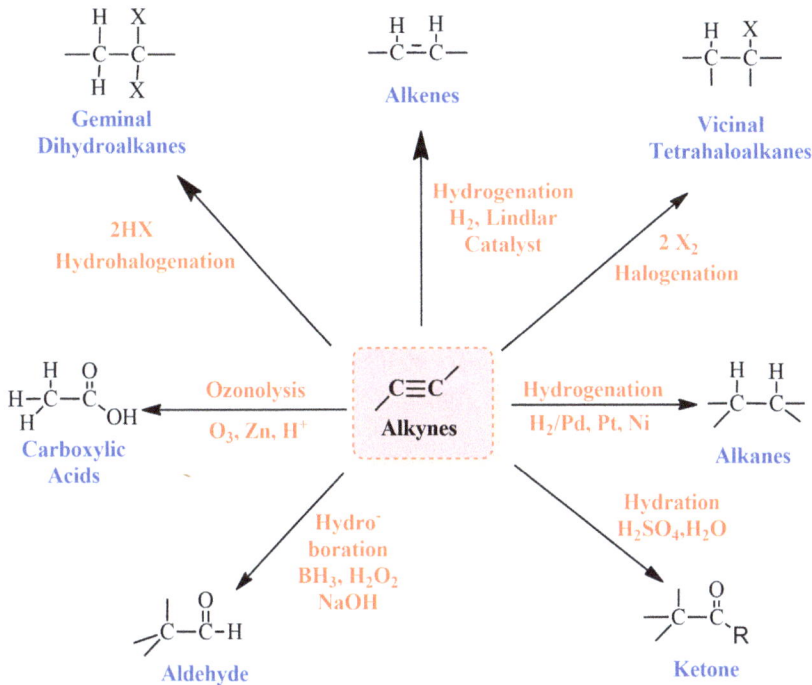

Figure 2.9: Reactions of dienes and alkynes.

2.8 Natural Gas and Petroleum

Natural gas is a colorless, highly flammable gas. Its primary constituents are ethane and methane. It is a kind of petroleum that can be found in crude oil. Natural gas is a fossil fuel and a natural resource that may be used to power vehicles, heat and cook food, and produce electricity. It is also used in the manufacture of polymers, dyes, and fertilizers. Natural gas is typically found in reservoirs dissolved in oil at high pressures, and it can also exist as a gas cap atop the oil. In many circumstances, the pressure of natural gas imposed on the underground oil resource provides the impetus to force oil to the surface.

This is known as associated gas; it is frequently referred to as the gaseous phase of crude oil and typically contains some light liquids such as propane and butane.

Petroleum, sometimes known as crude oil, is a type of fossil fuel. Petroleum was formed from the skeletal remains of ancient sea animals such as plants, algae, and bacteria, just like coal and natural gas. These organic remains (fossils) were changed into carbon-rich substances where we rely on as raw materials for fuel and a variety of products over millions of years of high heat and pressure.

Petroleum or crude oil is typically black or dark brown in color, but it can also be yellowish, reddish, tan, or greenish. Color variations represent the unique chemical compositions of various crude oil suppliers. Petroleum with little metals or sulfur, for example, is lighter and sometimes practically clear.

Petroleum is a complex mixture of hydrocarbons that must be refined into fractions before it can be used.

Refining (Figure 2.10) begins by distillation of crude oil into three principal cuts:
–   Straight-run gasoline/naphtha (C5–C11, bp 30–200 °C)
–   Kerosene (C11–C14, bp 175–300 °C)
–   Gas oil/diesel (C14–C25, bp 275–400 °C).

Finally, distillation under reduced pressure yields lubricating oils and waxes while leaving a non-distillable tarry residue of asphalt.

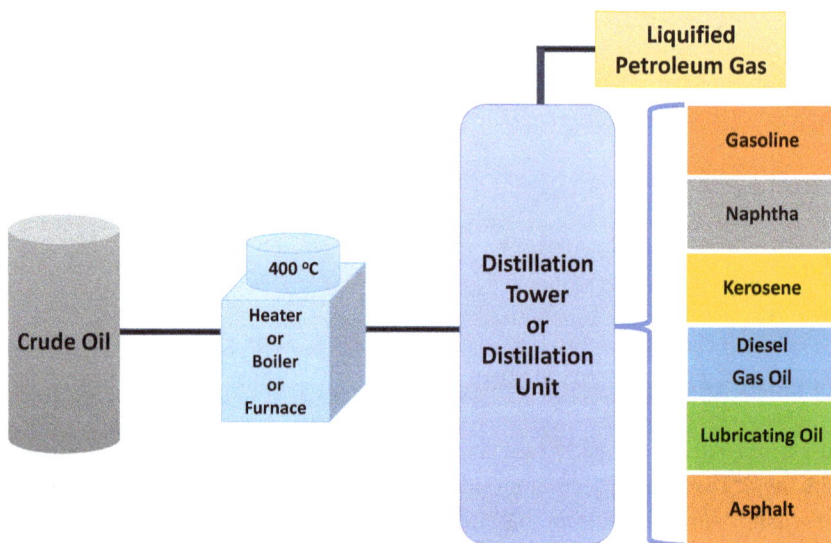

Figure 2.10: Basic crude oil refining process.

2.9 Polymers Made from Alkenes, Dienes, and Alkene Derivatives

A significant number of compounds containing carbon-carbon double bonds (Table 2.3) have been polymerized to produce useful materials.

Table 2.3: Polymers produced from compounds with carbon-carbon double bonds.

Monomer	Structure	Polymer (abbreviation)	Applications
Ethene	$CH_2=CH_2$	Polyethylene (PE) (HDPE) (LDPE)	Packaging material and squeeze bottles, food and beverage containers, and cleaning product bottles
Propene	$CH_2=CH–CH_3$	Polypropylene (PP)	Carpets and automobiles tires
Vinyl chloride	$CH_2=CH–Cl$	Poly (vinyl chloride) (PVC)	PVC tubes and pipes
Styrene	$CH_2=CH–C_6H_5$	Polystyrene (PS)	Housewares and radio and television cabinets
Acrylonitrile	$CH_2=CH–CN$	Polyacrylonitrile (PAN)	Ultra filtration membranes, fibers for textiles, and wool substitute for sweaters and blankets
1,1-Dichloroethene (Vinylidene Chloride)	$CH_2=CCl_2$	Polyvinylidene chloride or polyvinylidene dichloride (PVDC)	Water-based coating films
2-Methylpropene (isobutylene)	$CH_2=C(CH_3)_2$	Polyisobutene (polyisobutylene) (PIB)	Adhesives and sealants
Tetrafluoroethene	$CF_2=CF_2$	Teflon polytetrafluoroethylene (PTFE)	Nonstick coating for cookware
2-Chloro-1,3-butadiene (chloroprene)	$CH_2=C(Cl)CH=CH_2$	Neoprene (polychloroprene)	Electrical insulation and conveyer belts

2.10 Essential Terms

Alkanes Saturated hydrocarbons (have only C–C single bonds) with a general formula C_nH_{2n+2}.

Alkenes Unsaturated hydrocarbons which contain carbon-carbon double bond.

Alkyls Alkanes with one less hydrogen atom (alkane-H = alkyl).

Alkynes Hydrocarbons containing carbon-carbon triple bonds in addition to carbon-carbon single bonds and sometimes carbon-carbon double bonds.

Common names Also known as trivial names. Names that are based on historical roots with a specific name for each compound.

Combustion In air burning of alkanes in the natural gas, gasoline, and fuel oil to produce heat in addition to carbon dioxide and water.

Conjugated dienes Compounds that have the two double bonds separated by just one single bond.

Cumulated dienes Compounds that have the two double bonds are successive with no intervening single bonds.

Cycloalkanes Alicyclic compounds that can be represented by polygons in skeletal drawings.

Degree of alkyl substitution The number of alkyl groups bonded to a carbon atom in a compound.

Dehydration reaction The removal of H_2O (H and OH).

Dehydrohalogenation The removal of alkyl halides (H and X).

Dihalides A haloalkanes that have two halogens in their structure.

Ethyl group The CH_3CH_2-group.

Haloalkanes Organic compounds that have the general formula R-X.

Halogenations The chlorination or bromination of alkanes by treatment with chlorine or bromine in the presence of visible or UV light. It involves the addition of halogens (X_2).

Halogenated solvents Organic solvents with halogens in their structures. For example, CH_2Cl_2 and $CHCl_3$.

Hydration reaction The addition of water (H_2O, H, and OH).

Hydrocarbons Compounds that contain only carbon and hydrogen atoms in their structures.

Hydrogenation reaction The addition of hydrogens (H_2).

Isolated dienes Compounds that have the two double bonds separated with two or more single bonds.

IUPAC names Names that are based on rules set by International Union of Pure and Applied Chemistry.

Methyl group The CH3-group.

Nucleophilic aliphatic substitution A reaction that involves the attack by nucleophile on the substrate.

Pyrolysis The transformation of a compound into other compounds by heat alone without any oxygen.

Petroleum A complex mixture of hydrocarbons that must be refined into fractions before using.

Primary haloalkane A haloalkane in which halogen is attached to primary carbon.

Secondary haloalkane A haloalkane in which halogen is attached to secondary carbon.

Tertiary haloalkane A haloalkane in which halogen is attached to tertiary carbon.

2.11 Problems

2.11.1   Draw an alkane that has primary, secondary, tertiary, and quaternary carbons.

2.11.2   Draw a cycloalkane with the formula C_6H_{10}.

2.11.3   Draw the skeletal structure of 3-chloroheptane.

2.11.4   Write down the general formula of cycloalkenes. Give one example.

2.11.5   How many carbon and hydrogen atoms are there in the following structure?

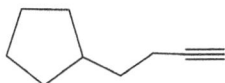

2.11.6   How many methylene groups are present in 2,3-dimethylnonane?

2.11.7   Provide the correct name for $CH_3CH_2CH(CH_3)_2$.

2.11.8   Draw the correct structure for 4-*tert*-butylnonane.

2.11.9   How many methyl groups are present in 2,4-dimetylheptane?

2.11.10 Provide the correct name for the $[(CH_3)_3C]_2CHCH_3$.

2.11.11 Predict all possible products for the following reactions:

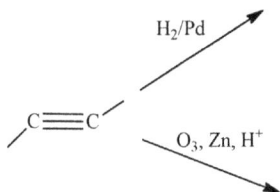

2.11.12 Circle or tick the reaction that follows Markovnikov's rule.

No.	Reactions		
1	R—C≡C—H $\xrightarrow[\text{Peroxide}]{\text{HX}}$ R—C=C—H (H, X)		
2	R—C≡C—H $\xrightarrow{H_2}$ R—C=C—H (H H, H H)		
3	R—C≡C—H $\xrightarrow{\text{HX}}$ R—C=C—H (X, H)		

2.11.13 What are the types of the following dienes?

2.11.14 Complete the following hydration reaction:

2.11.15 Write down the general formula for cycloalkenes. Give one example.

2.11.16 Draw the corresponding chemical structure for each of the following IUPAC names:

IUPAC name	Structure
5-Methyl-2-octyne	
4-Chloro-2-pentene	
1,3-Dibromocyclohexane	

2.11.17 Give the IUPAC name for $(CH_3)_2CHCH_2CH_2CH=CH_2$.

2.11.18 Give the IUPAC name for the following alkene derivative:

2.11.19 Draw the chemical structure for 4-chloropent-1-ene.

2.11.20 Draw the chemical structure for 1,2-dibromocyclohexene.

2.11.21 Draw the chemical structure for 8-methyl-2-nonen-6-yne.
for the internal one

2.11.22 Give one example for terminal alkyne and another example for the internal one.

2.11.23 What does Lindlar catalyst consist of?

2.11.24 Draw the chemical structure for 2-butyn-1-ol.

2.11.25 Show how enols rearrange to ketones.

2.11.26 Draw the chemical structure of benzyl chloride and explain why it is considered as substituted alkyl halide.

2.11.27 Draw the alcohols' chemical structures that are required for the synthesis of the following haloalkanes:

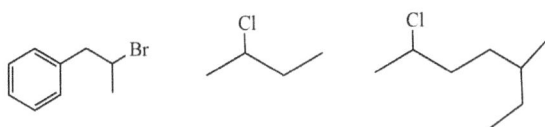

2.11.28 Give one example for nucleophilic aliphatic substitution reaction.
2.11.29 Show how the secondary haloalkane looks like.
2.11.30 Draw the structure of isopropyl bromide and define its class.

∗∗∗∗∗∗∗∗∗∗

Chapter 3
Aromatic Hydrocarbons

Objectives

After studying this chapter, learners will be able to:

- Write the common and IUPAC names of benzene derivatives and polynuclear aromatic compounds.
- Draw the chemical structures of benzene derivatives and polynuclear aromatic compounds.
- Describe the electrophilic aromatic substitution.
- Correlate the physical properties of benzene derivatives.
- Understand various preparation methods and reactions of benzene derivatives.
- Describe the uses and industrial applications of selected benzene derivatives.

3.1 Structure of Benzene

The molecular formula of benzene is C_6H_6, and its structure can be represented as shown in Figure 3.1. Benzene cannot be represented accurately by either individual Kekulé structure. The true structure is somewhere in between. All six hydrogens are equivalent. All carbon-carbon bonds in benzene have the same length, intermediate between typical single and double bonds. Benzene ring was also represented by other structures, but they are no longer in use.

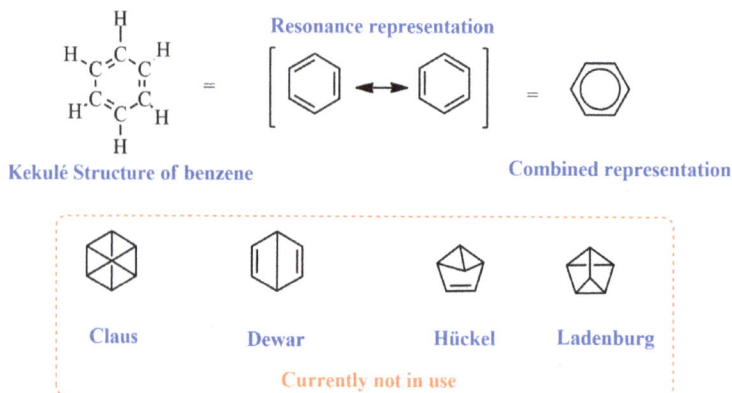

Figure 3.1: Structures of benzene ring.

https://doi.org/10.1515/9783111382753-003

3.2 Nomenclature (Naming) of Benzene and Benzene Derivatives

Many aromatic compounds that contain benzene rings in their structures are named by attaching the name of the substituents as a prefix to benzene. In some cases the substituent gives a special name to the ring and the aromatic compound in turn gets related common name. Examples of monosubstituted benzenes, C_6H_5Y, are shown in Figure 3.2.

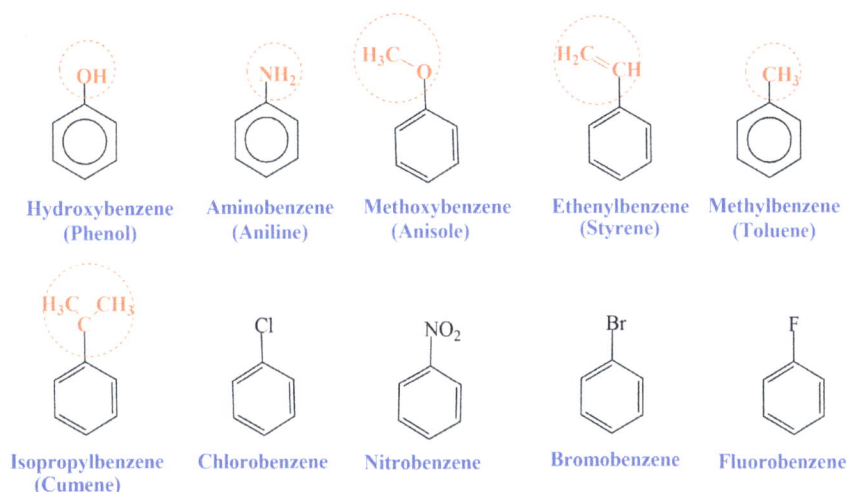

Hydroxybenzene (Phenol) Aminobenzene (Aniline) Methoxybenzene (Anisole) Ethenylbenzene (Styrene) Methylbenzene (Toluene)

Isopropylbenzene (Cumene) Chlorobenzene Nitrobenzene Bromobenzene Fluorobenzene

Figure 3.2: Examples of monosubstituted benzenes.

Different ways are used to name monosubstituted alkylbenzenes depending on the size of the alkyl group. If the alkyl substituent has six or fewer carbons, the arene is named as an alkyl-substituted benzene.

If the alkyl substituent has more than six carbons, the compound is named as a phenyl-substituted alkane. The phenyl (-**ph**) is used for the C_6H_5 unit when the benzene ring is considered as a substituent (Figure 3.3).

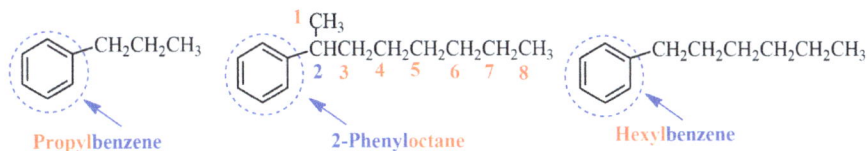

Propylbenzene 2-Phenyloctane Hexylbenzene

Figure 3.3: Alkyl-substituted benzenes.

Disubstituted benzenes are named using one of the prefixes *ortho* (*o*) **1,2**-relationship; *meta* (*m*) **1,3**-relationship; or *para* (*p*) **1,4**-relationship as shown in Figure 3.4.

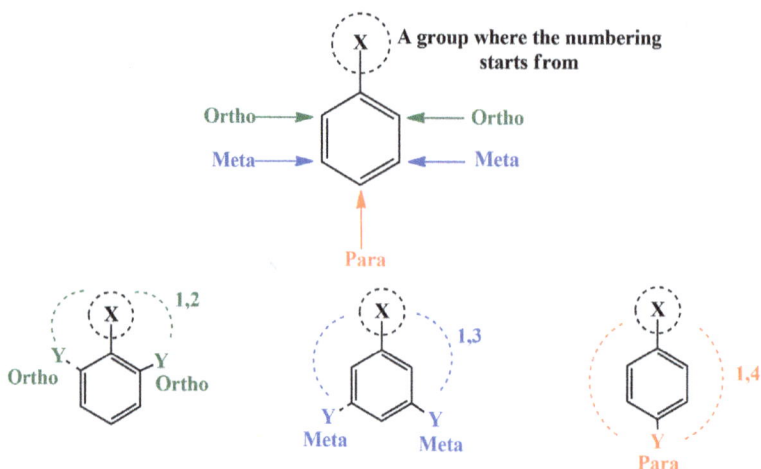

Figure 3.4: Ortho (*o*) 1,2-relationship; meta (*m*) 1,3-relationship; or para (*p*) 1,4-relationship.

When the two groups are different, and none of them gives a special name to the ring, the two groups are named in alphabetical order followed by the word benzene. In case one of the two groups gives a special name to the ring then the molecule is named as a derivative of that special name. When the two groups are the same and give a special name to the ring, then the molecule is named on the basis of these groups. All these examples are presented in Figure 3.5.

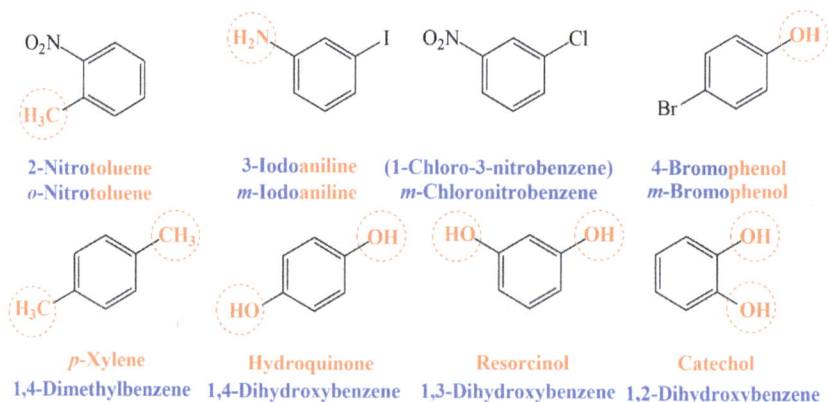

| 2-Nitrotoluene | 3-Iodoaniline | (1-Chloro-3-nitrobenzene) | 4-Bromophenol |
| *o*-Nitrotoluene | *m*-Iodoaniline | *m*-Chloronitrobenzene | *m*-Bromophenol |

| *p*-Xylene | Hydroquinone | Resorcinol | Catechol |
| 1,4-Dimethylbenzene | 1,4-Dihydroxybenzene | 1,3-Dihydroxybenzene | 1,2-Dihydroxybenzene |

Figure 3.5: Disubstituted benzenes with two different or similar groups.

If more than two groups are attached to the benzene ring, numbers are used to indicate their relative positions as shown in Figure 3.6.

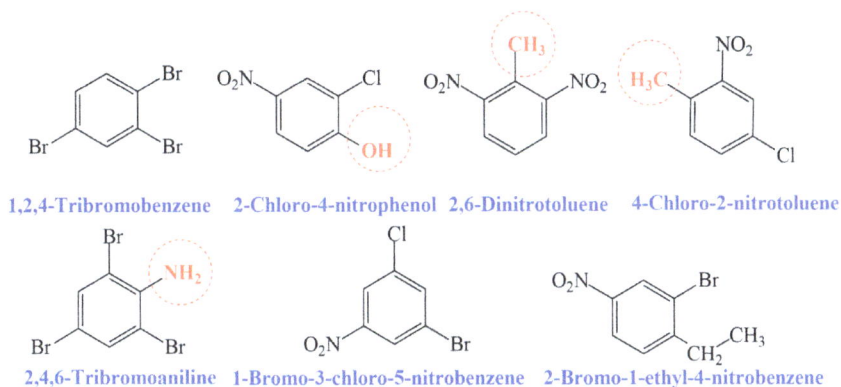

1,2,4-Tribromobenzene   2-Chloro-4-nitrophenol   2,6-Dinitrotoluene   4-Chloro-2-nitrotoluene

2,4,6-Tribromoaniline   1-Bromo-3-chloro-5-nitrobenzene   2-Bromo-1-ethyl-4-nitrobenzene

Figure 3.6: Benzenes with more than two groups.

3.3 Polynuclear Aromatic Hydrocarbons

Polynuclear aromatic hydrocarbons are molecules that have fused aromatic rings and share a pair of carbon atoms in their structures. Naphthalene, anthracene, phenanthrene, and pyrene are examples of this type of molecules (Figure 3.7).

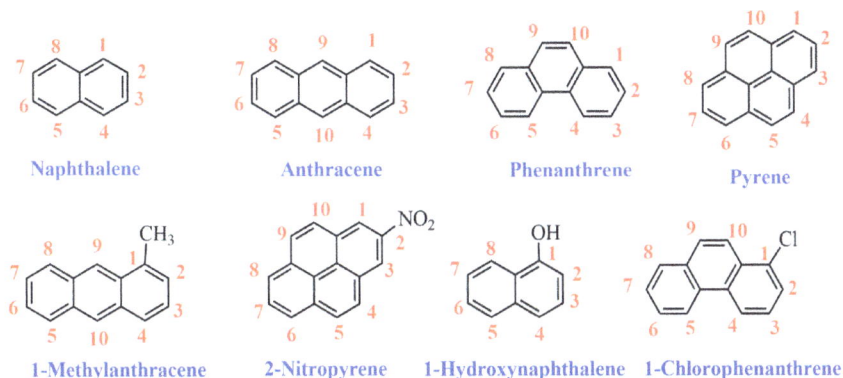

Naphthalene   Anthracene   Phenanthrene   Pyrene

1-Methylanthracene   2-Nitropyrene   1-Hydroxynaphthalene   1-Chlorophenanthrene

Figure 3.7: Polynuclear-fused aromatic rings and selected monosubstituted examples.

3.4 Electrophilic Aromatic Substitution (Benzene and Benzene Derivatives)

Benzene ring and benzene derivatives undergo a series of electrophilic substitution reactions to give different substituted products as a result of an attack by different electrophiles. In these reactions, hydrogen is replaced by different groups such as R,

X, NO_2, COR, and SO_3H. The general mechanism of electrophilic aromatic substitution with different electrophiles is summarized in Figure 3.8.

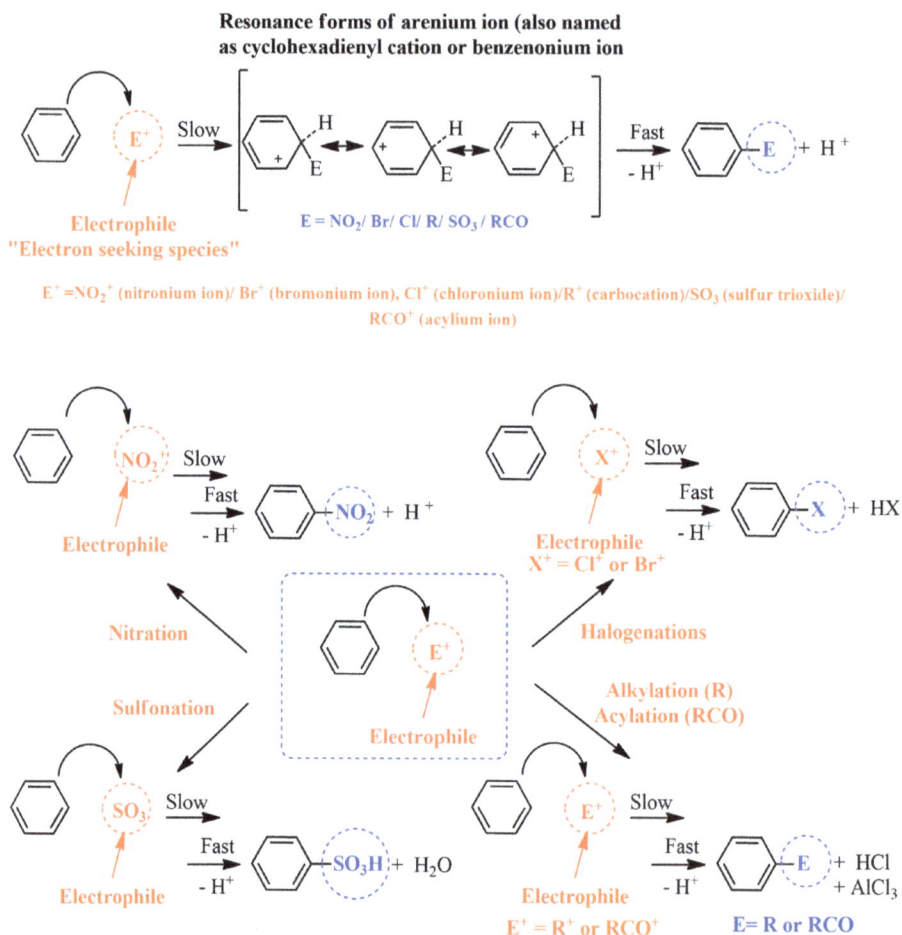

Figure 3.8: General mechanism of electrophilic aromatic substitution with different electrophiles.

3.5 Physical Properties of Benzene and Benzene Derivatives

Benzene is a "carcinogenic" colorless liquid with boiling point around 80 °C. It is not miscible with water but soluble in organic solvents. It is less dense than water with a density of 0.874 g/mL at 25 °C.

Halobenzenes on the other hand are colorless compounds. Their boiling points are to some extent similar to the corresponding haloalkanes. For example, chlorobenzene (bp = 132 °C) and bromobenzene (bp = 156 °C) have boiling points very nearly the same as those of *n*-hexyl chloride (bp = 134 °C) and *n*-hexyl bromide (bp = 156 °C). They are insoluble in water, denser than water, and soluble in organic solvents. Simple hydroxybenzenes are liquids or low-melting solids and have quite high boiling points. Hydroxybenzene (phenol) itself is somewhat soluble in water (9 g per 100 g of water), but most other hydroxybenzenes are insoluble in water. Unless some group capable of producing color, hydroxybenzenes themselves are colorless. They are fairly acidic compounds.

3.6 Preparation and Reactions of Benzene Derivatives

When electrophilic substitution takes place on benzene ring only one product is formed but when the same reaction is carried out on benzene ring that has a substituent then the story is different. Substituents have an effect on the reactivity of benzene ring, and some activate the ring while others deactivate the ring. Substituents also have an effect on the orientation of the reaction on the ring. So the type and nature of the substituent determines the position of the second substitution. Substituents can be classified as ortho- and para-directing activators, ortho- and para-directing deactivators, and meta-directing deactivators. Table 3.1 lists these three classes.

Table 3.1: The effect of substituent (in monosubstituted benzene ring) in electrophilic aromatic substitution.

Substituent	Orientation	Reactivity
$-CH_3$	Ortho- and para-directing	Activating
$-OH$	Ortho- and para-directing	Activating
$-NH_2$		
$-Cl$	Ortho- and para-directing	Deactivating
$-Br$		
$-F$		
$-I$		
$-NO_2$	Meta-directing	Deactivating
$-CN$		
$-CHO$		
$-COCH_3$		
$-CO_2CH_3$		

3.6.1 Halobenzenes (Aryl Halides)

Preparation of halobenzenes by direct halogenation of the aromatic ring is more use-ful than direct halogenation of alkanes. The final product obtained will depend on the substitution directing groups and the attack is not nearly as random as in the free radical halogenation of hydrocarbons. Halobenzenes are most commonly prepared by replacement of the nitrogen of a diazonium salt (Figure 3.9).

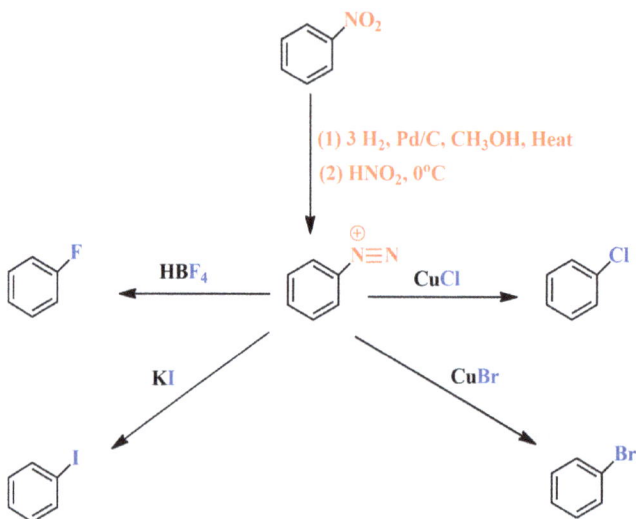

Figure 3.9: Preparation of halobenzenes.

On the other hand, halobenzenes react with magnesium metal to form aryl Grignard reagents. In addition, halobenzenes can undergo the typical electrophilic aromatic substitution reactions such as nitration, Friedel-Crafts alkylation, sulfonation, and hal-ogenations. All above-mentioned reactions are presented in Figure 3.10.

3.6.2 Hydroxybenzenes (Phenols)

Among the hydroxybenzenes, phenol is the most important member of the family. A certain amount of phenol, as well as the cresols, is obtained from coal tar, but nearly most of the phenols are made by chemical means. A fusion of sodium benzenesulfo-nate with alkali is one of the methods for making alkyl-substituted phenols. Another way of making phenol is by using the Dow process that involves the reaction of chlo-robenzene with aqueous NaOH at 360 °C followed by acid treatment. Nearly all phenol is made today, however, by a newer process that starts with cumene "isopropyl ben-

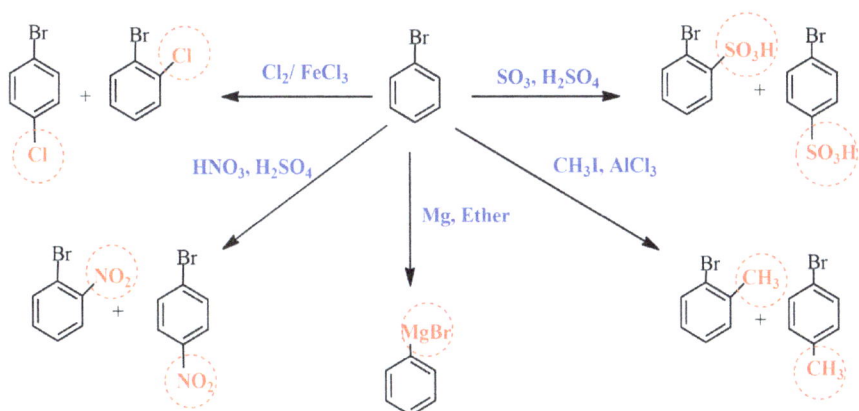

Figure 3.10: Reactions of halobenzenes.

zene." Cumene is converted by air (oxidation) into cumene hydroperoxide, which is converted by aqueous acid into phenol and acetone (Figure 3.11).

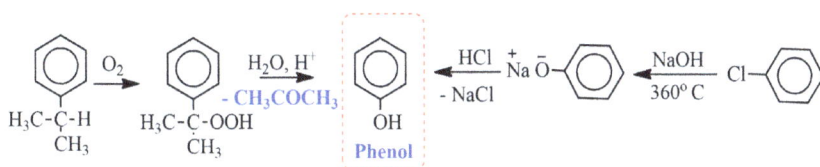

Figure 3.11: Preparation of phenol.

On the other hand, phenol has high reactivity toward electrophilic substitution. It undergoes not only those electrophilic substitution reactions but also many others that are possible only because of the unusual reactivity of the ring. The –OH group activates some positions around the ring more than others, so the incoming groups will prefer to go into some positions much faster than they will into others. It has a 2,4-directing effect which is also known as an ortho-para effect. For example, phenol reacts with dilute nitric acid at room temperature to give a mixture of 2-nitrophenol and 4-nitrophenol and also reacts with concentrated nitric acid to give 2,4,6-trinitrophenol (picric acid). The reaction of phenol with concentrated H_2SO_4 at room temperature gives o-phenol sulfonic acid due to the formation of hydrogen bonds between hydroxyl and sulfonyl groups.

At high temperatures, no such interaction between the two groups because the steric repulsion overcomes the attraction, and therefore, the para product is obtained. Phenol can also react with bromine water to form 2,4,6-tribromophenol as shown in Figure 3.12.

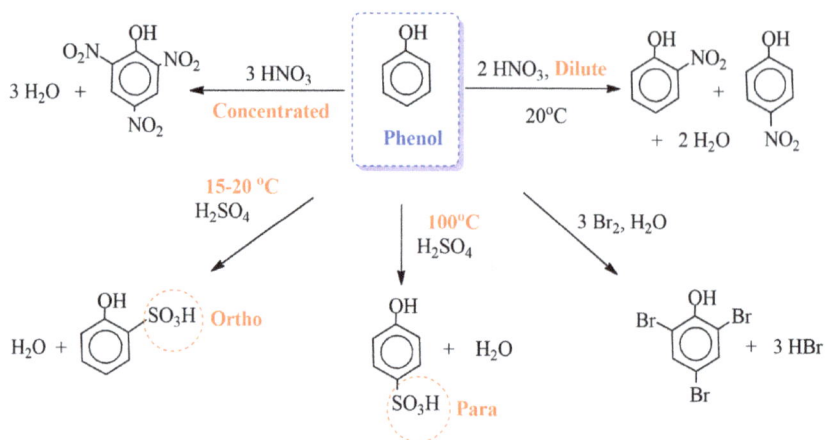

Figure 3.12: Reactions of phenol.

3.7 Industrial Applications of Benzene and Benzene Derivatives

Selected industrial applications for benzene and benzene derivatives are listed in Table 3.2.

Table 3.2: Industrial applications of benzene derivatives.

Name	Commercial use
Benzene	Used to make lubricants, rubbers, detergents, drugs, pesticides, and dyes
Benzaldehyde	Used chiefly as precursor for other organic compounds ranging from pharmaceuticals to plastic additives
Benzoic acid	Its salts and esters are used as raw materials for pharmaceutical applications, food, industrial preservatives, plasticizers, and fibers
Toluene	Used as a solvent for paints and coatings and also used as adhesive solvent in plastic toys
Styrene	Used as a basic component of polystyrene and styrofoam plastics
Phenol	Used in medical antiseptics and bactericides such as Dettol and also as a component of epoxy resins
Aniline	Used in the production of rubber accelerators and in the manufacture of azodyes
Chlorobenzenes	Used extensively as insecticides, herbicides, fungicides, and bactericides

3.8 Essential Terms

Anthracene A fused aromatic compound formed from three fused benzene rings.

Benzene Aromatic ring with general formula C_6H_6.

Catechol Ortho-dihydroxybenzene.

Cresol Methyl phenol.

Diazonium salt Organic compound that has the structure ArN_2^+.

Disubstituted benzenes Benzene ring with one group attached to it.

Electrophilic aromatic substitution reaction A reaction that takes place in two steps reaction of an electrophile, E^+, with the aromatic ring, followed by loss of H from the carbocation intermediate to the substituted aromatic ring.

Grignard reagent A compound that has the structures R-Mg-X or Ar-Mg-X.

Haloalkanes Organic compounds that have the general formula Ar-X.

Hydroquinone Para-dihydroxybenzene.

Meta A 1,3-relationship on a benzene ring.

Monosubstituted benzene Benzene ring with one group attached to it.

Naphthalene A fused aromatic compound formed from two fused benzene rings.

Ortho A 1,2-relationship on a benzene ring.

Para A 1,4-relationship on a benzene ring.

Phenanthrene A fused aromatic compound formed from three fused benzene rings.

Phenols Compounds of the general formula "ArOH."

Phenyl group (Ph or C_6H_5) The benzene ring when named as a substituent or a group attached to another molecule.

Polynuclear aromatic hydrocarbons Aromatic rings that share a pair of carbon atoms, so they are fused.

Resorcinol Meta-dihydroxybenzene.

3.9 Problems

3.9.1   Draw the chemical structures of the following aromatic compounds:
   i.   *p*-Bromoanisole
   ii.   2-Methoxyphenol
   iii. *m*-Chlorobenzoic acid
   iv.   3-Chloroaniline
   v.   *o*-Ethyltoluene

3.9.2   Benzene undergoes chlorination reaction. Name the electrophile involved in this reaction.

3.9.3   Draw the chemical structure of naphthalene and assign numbers to the carbon atoms.

3.9.4   Name the following aromatic compound:

3.9.5   Which of the following phenols is 3-nitrophenol?
         i.   o-Nitrophenol
         ii.   m-Nitrophenol
         iii. p-Nitrophenol

3.9.6   Name the following toluene derivative:

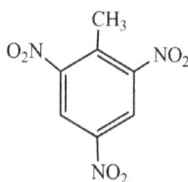

3.9.7   What are the IUPAC names of the following benzene derivatives?
         i.   Phenol
         ii.   Toluene
         iii. Aniline

3.9.8   Write the name of each of the following electrophiles:
         i. Cl^+
         ii. R^+
         iii. RCO^+

3.9.9   List down two industrial applications for styrene.

3.9.10   What is the effect of CH_3 substituent in electrophilic aromatic substitution?

3.9.11   What is resorcinol?

3.9.12   Assign numbers to the carbon atoms of the following polynuclear aromatic compounds:

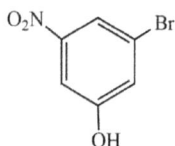

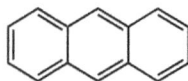

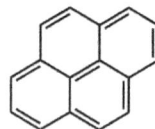

3.9.13   Describe the alkylation reaction of benzene.

3.9.14   What is the final product for the sulfonation of benzene at 100 °C?

3.9.15   List two industrial applications for halobenzenes.

3.9.16   Insert nitro groups at ortho- and para-positions in the following structure:

OH

3.9.17   Draw the chemical structure of catechol.

3.9.18   Give the IUPAC name for cumene.

3.9.19   What is the rule of FeX_3 in the halogenation reaction of benzene?

3.9.20   What kind of reaction is the sulfonation of benzene?

3.9.21   Write the general formula of haloalkanes.

3.9.22   Draw the chemical structure of 1-ethyl-2-nitrobenzene.

3.9.23   Explain why aromatic nucleophilic substitution is not preferable for haloalkanes to undergo.

3.9.24   Draw the general structure of Grignard reagent and give an example.

3.9.25   Complete the missing information in the following reactions:

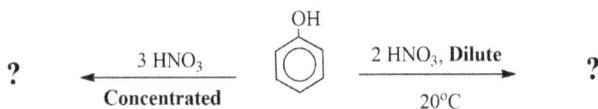

OH

? $\xleftarrow[\text{Concentrated}]{3\ HNO_3}$ $\xrightarrow[20°C]{2\ HNO_3,\ \textbf{Dilute}}$?

3.9.26   Describe the complete reaction for the nitration of bromobenzene.

3.9.27   Describe the complete reaction for the alkylation of bromobenzene.

3.9.28   Draw the chemical structure for *m*-dimethoxybenzene.

3.9.29   What are the main differences between catechol, resorcinol, and hydroquinone?

3.9.30   Predict the major product of the following reaction:

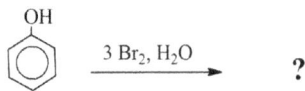

OH

$\xrightarrow{3\ Br_2,\ H_2O}$?

✶✶✶✶✶✶✶✶✶✶✶✶

Chapter 4
Amines, Alcohols, Thiols, Ethers, Sulfides, and Heterocyclic Compounds Containing Nitrogen, Oxygen, and Sulfur

Objectives

After studying this chapter, learners will be able to
- Write their common and IUPAC names.
- Draw their chemical structures.
- Describe important methods for their preparation and briefly discuss their reactions.
- Correlate their physical properties.

4.1 Amines

Both alkyl and arylamines are derived from ammonia. In the alkylamines, only alkyl groups are attached to the nitrogen atoms, while in the aryl amines the structure can have aryl groups only or both aryl and alkyl groups are attached to the nitrogen atoms. Amines can be classified into primary (1°), secondary (2°), or tertiary (3°) depending on how many hydrogen atoms are substituted by R or Ar groups.

Amines are often named by simply designating the groups attached to nitrogen followed by the suffix amine or by the prefix amino when it is considered as a substituent and attached to other parent compounds. The prefix di- or tri- is added only to the alkyl or aryl groups of the symmetrical (same groups attached to nitrogen atom) secondary and tertiary amines. Asymmetrical (different groups attached to nitrogen atom) secondary and tertiary amines are named N-substituted amines.

The largest alkyl group is chosen as the parent name, and other alkyl groups are considered N-substituents on the parent. Selected examples of the above-mentioned amines are presented in Figure 4.1 and Table 4.1.

4.1.1 Physical Properties of Amines

Low-molecular weight primary, secondary, and tertiary amines dissolve in water due to their ability to form hydrogen bonds with water. Primary and secondary amines can form hydrogen bonds with each other while tertiary amines cannot form hydrogen bonds with each other and due to that, tertiary amines have lower boiling points than primary and secondary amines of comparable molecular weights.

https://doi.org/10.1515/9783111382753-004

Table 4.1: Selected examples of amines.

Primary (**1°**), secondary (**2°**), and tertiary (**3°**) alkyl and aryl amines	
Symmetrical secondary and tertiary amines (*same groups attached to nitrogen atom*)	
Asymmetrical secondary and tertiary amines (*different groups attached to nitrogen atom*)	

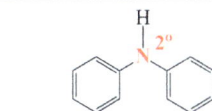

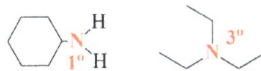

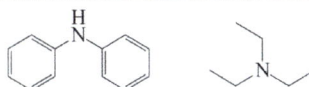

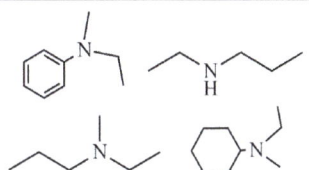

Ethylamine
(**Alkyl**)

Aniline
(**Aryl**)

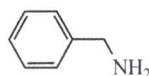

Benzylamine
(**Substituted alkyl**)

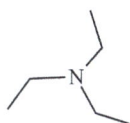

Triethylamine

Dimethylamine

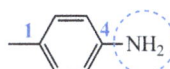

4-Aminotoluene

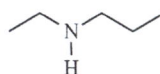

N,N-**Dimethylpropylamine**   *N*-**Ethyl-***N*-**methylcyclohexylamine**   *N*-**Ethylpropylamine**

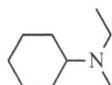

Figure 4.1: Naming of selected amines.

4.1.2 Preparation of Amines

Alkyl azides are reduced to primary alkylamines by lithium aluminum hydride or by catalytic hydrogenation. Nitriles are reduced to primary alkylamines by lithium aluminum hydride.

Lithium aluminum hydride reduces the carbonyl group of the amide to the methylene group. Primary, secondary, or tertiary amines may be prepared by proper choice of the starting amide. Water is used to quench the reaction after completion in all the above-mentioned reactions. These reactions for the preparation of amines are summarized in Figure 4.2.

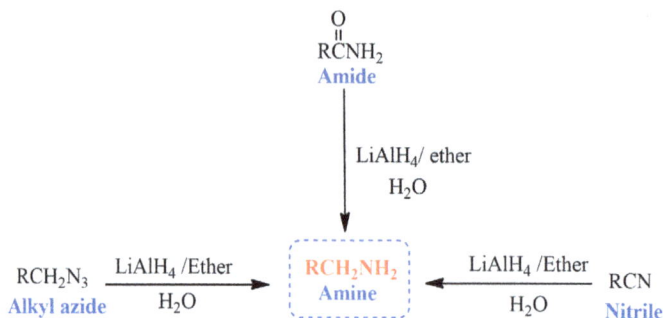

Figure 4.2: Preparation of amines.

4.1.3 Industrial Sources

Simple methylated amines are prepared by the reaction of ammonia with methanol in the presence of an alumina catalyst. The reaction yields a mixture of *mono-*, *di-*, and *tri-*methylated products that can be easily separated by distillation.

Aniline, the most important of all amines, can be prepared by the reduction of nitrobenzene with iron, zinc, tin, and tin (II) chloride ($SnCl_2$) in dilute hydrochloric acid or by catalytic hydrogenation. Treatment of chlorobenzene with ammonia in the presence of a catalyst can also result in the formation of aniline.

4.1.4 Reactions of Amines

Amines can be converted into their salts by aqueous mineral acids and can be liberated back from their salts by aqueous hydroxides. Amines also react with alkyl halides to produce amines of the next higher class. Primary and secondary amines also react with acid chlorides to form *N*-substituted amides, while tertiary amines fail to do so because they do not have hydrogen to lose to stabilize the product. The above-mentioned reactions are presented in Figure 4.3.

$$\overset{\overset{\displaystyle O}{\underset{\displaystyle \|}{}}}{RCNHR}$$

RCNHR
Amide

↑

RCOCl

| ⊕ ⊖ RNH₃Cl Ammonium Chloride | ⟵ HCl ⎯⎯ NaOH ⟶ | RNH₂ Amine | ⎯RX⟶ ⎯RX⟶ ⎯RX⟶ | ⊕⊖ R₄NX Quaternary Ammonium Salt |

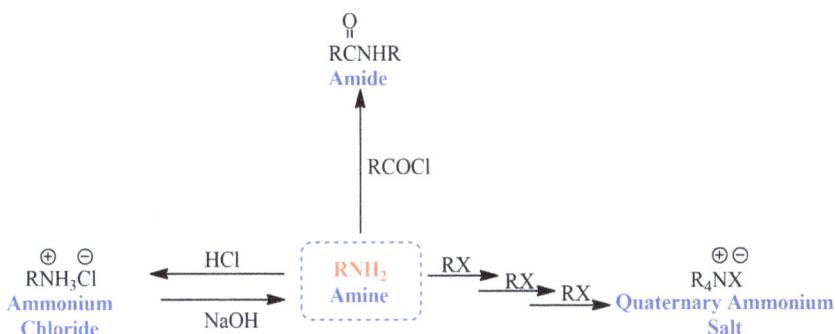

Figure 4.3: Reactions of amines.

4.2 Alcohols

4.2.1 Structure and Nomenclature of Alcohols

Alcohols are compounds that have hydroxyl groups bonded to saturated sp^3-hybridized carbon atoms. They are classified as primary (1°), secondary (2°), or tertiary (3°), based on the number of alkyl groups attached to the hydroxyl-bearing carbon.

According to the IUPAC naming system, simple alcohols are named as derivatives of the parent alkane using the suffix -ol.

To name alcohols first derive the parent name by replacing the -e ending with -ol in the alkane of the longest continuous carbon chain. Then start the numbering at the end nearest to the hydroxyl group and finally number the substituents according to their position in the chain and write the name listing the substituents in alphabetical order. Examples are shown in Figure 4.4.

4.2.2 Physical Properties of Alcohols

Simpler alcohols such as methanol and ethanol are liquids at room temperature and pressure, volatile, and colorless and have a characteristic smell. They mix with water in all proportions because of their hydrophilic region. Their borderline of solubility occurs at four to five carbons for 1 alcohol. Some alcohols are known with a carbon chain of about 20 atoms. These are solids resembling paraffin wax in appearance.

Alcohols' boiling points are higher than that of the alkanes with the same number of carbon atoms because of their hydrogen bonding (Figure 4.5).

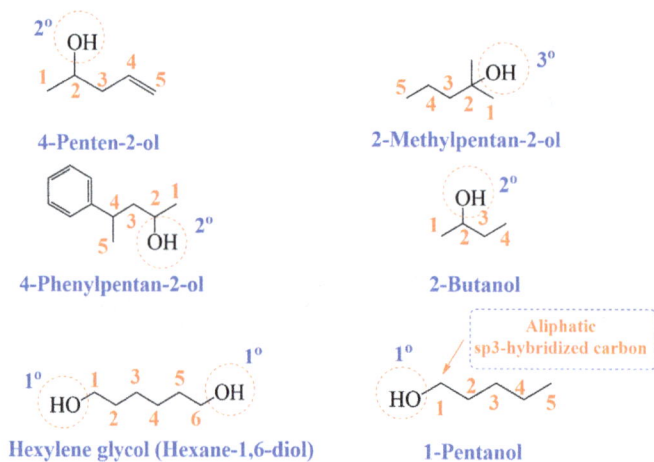

4-Penten-2-ol

2-Methylpentan-2-ol

4-Phenylpentan-2-ol

2-Butanol

Hexylene glycol (Hexane-1,6-diol)

1-Pentanol

Figure 4.4: Selected examples of alcohols.

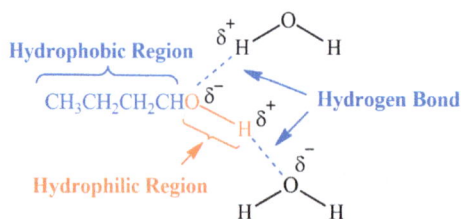

Figure 4.5: Hydrogen bonding in alcohols.

4.2.3 Preparation of Alcohols

Alcohols can be prepared from alkenes by either direct hydration with water in the presence of a catalyst or indirect hydration by the addition of sulfuric acid to alkane followed by hydrolysis of the alkyl hydrogen sulfate. They can also be prepared by hydroboration that involves treating alkenes with diborane to give alkyl boranes, R_3B, on oxidation with alkaline hydrogen peroxide gives alcohol. This method always leads to anti-Markovnikov's addition of water to alkenes.

In addition to the above-mentioned methods they can also be prepared by boiling alkyl halides with an aqueous solution of alkali hydroxide. This general procedure produces primary and secondary alcohols. All above-mentioned reactions are summarized in Figure 4.6.

$$H_2C=CH_2 \xrightarrow[\text{H}_2\text{O, H}^+]{\text{Hydration}} \underset{CH_2\ CH_2}{\overset{H\ \ OH}{|\ \ |}} \xleftarrow[\text{KOH, -KX}]{\text{Hydrolysis}} \underset{CH_2\ CH_2}{\overset{H\ \ X}{|\ \ |}}$$

$$H_2C=CH_2 \xrightarrow[\text{2H}_2,\ \text{CO, Cobalt}]{\text{Hydroformylation}} \underset{CH_2\ CH_2}{\overset{CH_3\ OH}{|\ \ |}} \xleftarrow[(BH_3)_2,\ H_2O_2,\ OH^{\ominus}]{\text{Hydroboration}} \underset{CH:CH_2}{\overset{CH_3}{|}}$$

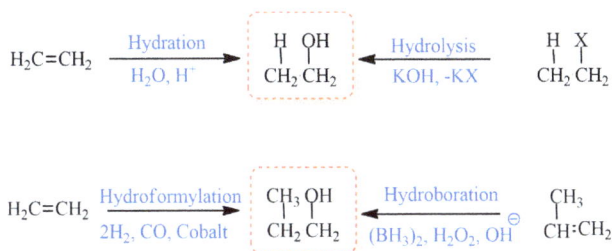

Figure 4.6: Preparation of alcohols.

4.2.4 Reactions of Alcohols

Reactions of alcohols take place at the C–O bond when alcohols are subjected to acid-catalyzed dehydration or converted to alkyl halides and at the C–H bond when primary alcohols are oxidized to aldehydes or carboxylic acids and when secondary alcohols are oxidized to ketones.

Reactions also take place at the O–H bond when alcohols are converted to alkoxides or ethers. All these reactions are presented in Figure 4.7, and the summary of the reactions is shown in Figure 4.8.

4.2.5 Polyhydroxy Alcohols

Polyhydroxy alcohols are compounds that have more than one hydroxyl group attached to different carbons of the structure. These compounds can be dihydroxy with two OH-groups such as 1,2-ethanediol(ethylene glycol) or trihydroxy with three OH-groups such as 1,2,3,-propanetriol (glycerol). Sugar alcohols can also be considered polyhydroxy alcohols. Figure 4.9 shows examples of polyhydroxy alcohols.

4.3 Thiols (Mercaptans)

4.3.1 Structure and Nomenclature of Thiols

Mercaptans or thiols (R-SH) are sulfur analogs of alcohols (R-OH). They are organosulfur compounds consisting of carbon, hydrogen, and sulfur. They are named by adding the suffix -thiol to the name of the corresponding alkane, and numbering the chain starts from the carbon that bears the –SH group. A mercapto group or a sulfhydryl group is given to the –SH when it exists as a substituent (Figure 4.10).

O-H Reactions
Formation of
alkoxides and ethers

$$RCH_2OH \xrightarrow{NaNH_2} RCH_2O^{\ominus} Na^{\oplus} + NH_3$$

$$RCH_2OH \xrightarrow{NaH} RCH_2O^{\ominus} Na^{\oplus} + H_2$$

$$RCH_2OH \xrightarrow{H_2SO_4, 140^\circ C} RCH_2OCH_2R + H_2O$$

C-O Reactions
Formation of alkenes
and alkyl halides

C-H Reactions
Formation of aldehydes or
ketones or carboxylic acid

$$R-\overset{R(H)}{\underset{H}{C}}-OH \xrightarrow{SOCl_2} R-\overset{R(H)}{\underset{H}{C}}-Cl + SO_2 + HCl$$

$$-\overset{H}{\underset{H}{C}}-\overset{OH}{\underset{H}{C}}- \xrightarrow{H_2SO_4, 180^\circ C} -\overset{}{\underset{H}{C}}=\overset{}{\underset{H}{C}}- + H_2O$$

$$RCH_2OH \xrightarrow[CH_2Cl_2]{PCC\ or\ PDC} RCHO$$

$$RCH_2OH \xrightarrow[H_2SO_4, H_2O]{K_2Cr_2O_7} RCOOH$$

$$R_2CHOH \xrightarrow[H_2SO_4, H_2O]{K_2Cr_2O_7} RCOR$$

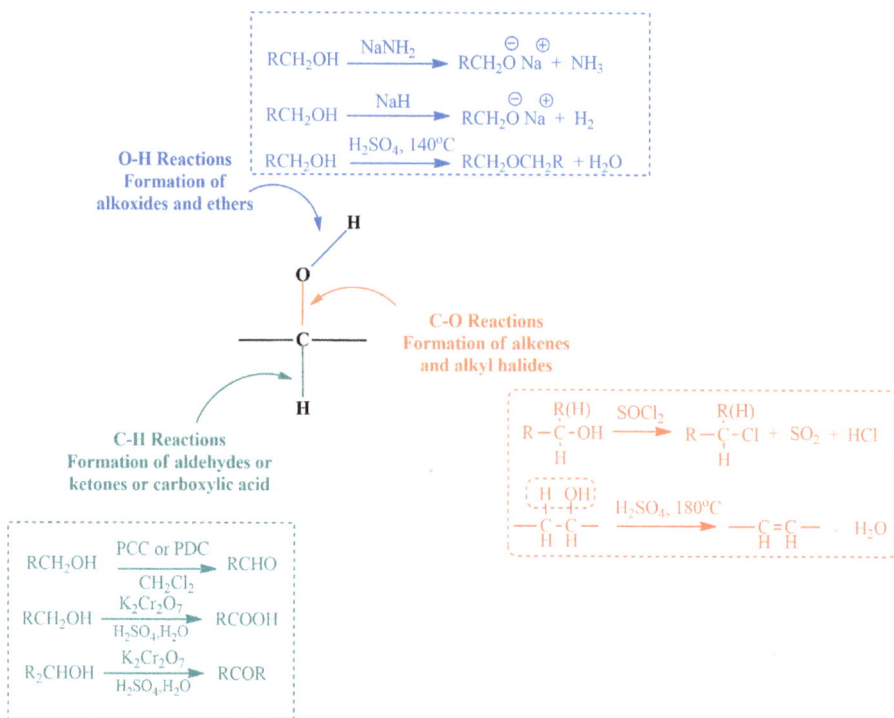

Figure 4.7: Reactions of alcohols.

4.3.2 Physical Properties of Thiols

Low-molecular weight thiols have a strong-smelling odor, and it weakens with the increasing number of carbons due to the decrease in both the volatility and the sulfur content. The S–H bond is less polar than the O–H bond and hydrogen bonding in thiols is much lower than that of alcohols. For example, ethanethiol has a boiling point of around 35 °C and is soluble in water to the extent of 0.7 g/100 mL (20 °C) while ethanol has a boiling point of around 75 °C and is very soluble in water. Another example is 1-butanol which has a boiling point of around 118 °C and is soluble in water to the extent of 7.3 g/100 mL (25 °C) compared to 1-butanethiol that has a boiling point around 98.2 °C and soluble in water to the extent of 0.06 g/100 mL (25 °C).

4.3.3 Preparation and Reactions of Thiols

Thiols can be prepared from alkyl halides by S_N2 displacement with a hydrosulfide ion. Thiols react with very mild oxidizers like Br_2 or I_2 to form disulfides. The disul-

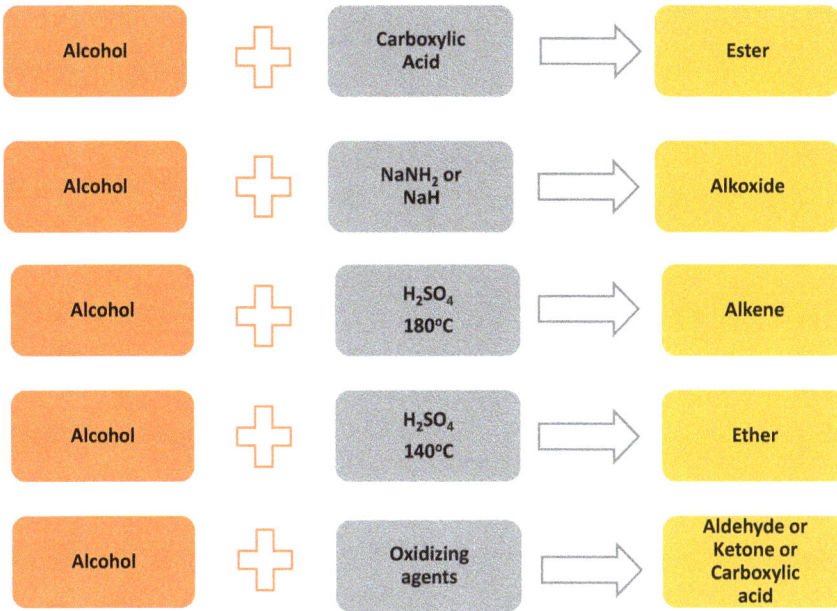

Figure 4.8: Summary of alcohols reactions.

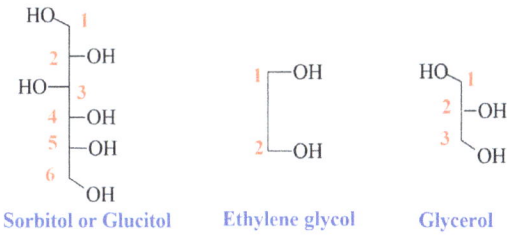

Sorbitol or Glucitol   Ethylene glycol   Glycerol

Figure 4.9: Example of polyhydroxy alcohols.

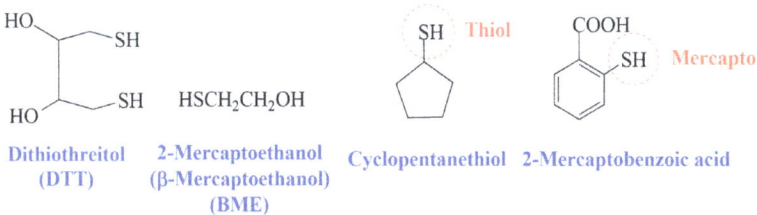

Dithiothreitol   2-Mercaptoethanol   Cyclopentanethiol   2-Mercaptobenzoic acid
(DTT)   (β-Mercaptoethanol)
(BME)

Figure 4.10: Selected examples of thiols.

fides can be reduced back to thiols by treatment with zinc and acid mixture or by either dithiothreitol (DTT) or β-mercaptoethanol (BME).

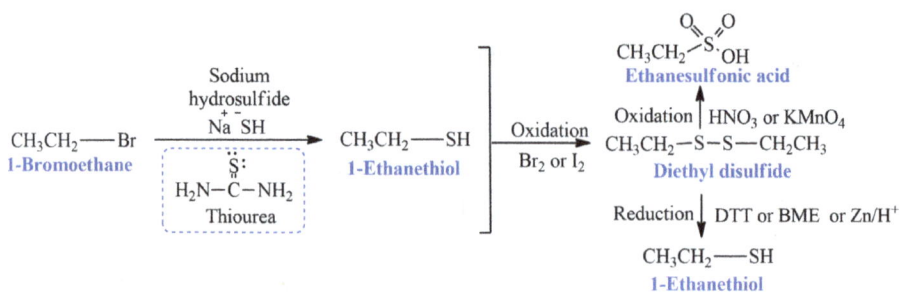

Figure 4.11: Preparation and reactions of thiols.

The interconversion between disulfides and thiols is a redox reaction (reduction-oxidation type reaction). Disulfides can be further oxidized to sulfonic acid by stronger oxidizers like HNO_3 and $KMnO_4$ as shown in Figure 4.11.

4.4 Ethers

4.4.1 Structure and Nomenclature of Ethers

Ethers are a class of organic compounds that have an oxygen atom bonded to two alkyl or aryl groups. They have a general formula that can be presented as R-O-R', Ar-O-R, and Ar-O-Ar, where R and R' are any alkyl groups and Ar is phenyl or any other aromatic group. They are named by listing the two substituents or groups linked to the oxygen atom followed by the word ether. Ethers can be symmetrical when the two groups are identical such as diethyl ether and unsymmetrical if the two groups are different such as methyl phenyl ether.

If other functional groups are present in the same structure, the ether part is considered as an alkoxy substituent such as methoxy or ethoxy. There are also cyclic and crown ethers. Selected examples are shown in Figure 4.12.

4.4.2 Physical Properties of Ethers

Ethers don't show any intermolecular hydrogen bonding between their molecules because the oxygen atom in ether has no hydrogen attached to it. Due to that fact, ether has boiling points almost the same as alkanes of comparable molar mass and much lower than alcohols with comparable molar mass.

For example the boiling point of dimethyl ether (molar mass = 46) is −24.8 °C compared to propane −42 °C (molar mass = 44) and ethyl alcohol 78 °C (molar mass = 46).

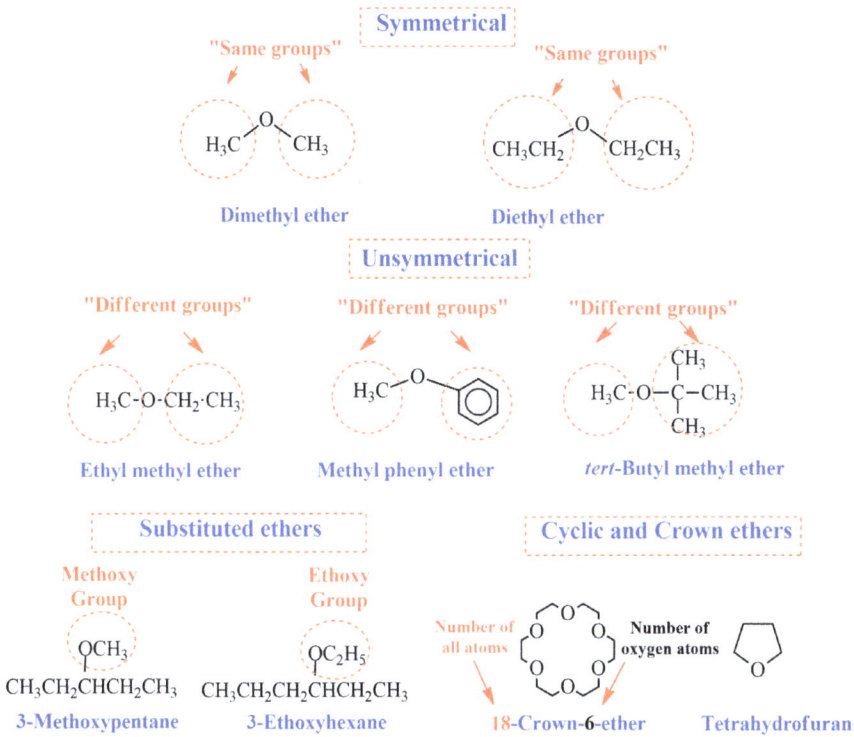

Figure 4.12: Selected examples of ethers.

On the other hand, ethers show solubility range in water comparable to that of the alcohols, both diethyl ether and n-butyl alcohol, for example, being soluble to the extent of about 6–7 g per 100 g water. The water solubility of ethers is due to the hydrogen bonding of the ether's oxygen and water hydrogens (Figure 4.13).

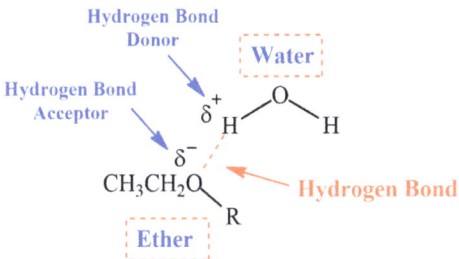

Figure 4.13: Hydrogen bonding scheme in ethers.

4.4.3 Preparation of Ethers

Symmetrical and asymmetrical ethers are generally prepared by Williamson synthesis. It involves a nucleophilic displacement of a halide ion by an alkoxide or phenoxide ion. They are also prepared by reacting an alcohol with sulfuric acid at 140 °C. Another method of preparing ether is by reacting alkene with an alcohol in the presence of mercuric acetate (alkoxymercuration) followed by reduction with $NaBH_4$ (demercuration) (Figure 4.14).

Figure 4.14: Preparation of ethers.

4.4.4 Reactions of Ethers

Ethers are unreactive to many reagents used in organic chemistry, and due to that property ethers are mostly used as reaction solvents. They are cleaved only by strong acids such as aqueous hydroiodic acid or concentrated hydrobromic acid. In the presence of oxygen, ethers slowly autoxidize to form hydroperoxides and dialkyl peroxides. If concentrated or heated, these peroxides may explode. Examples for these reactions are shown in Figure 4.15.

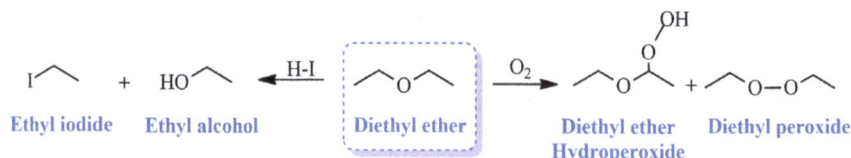

Figure 4.15: Reactions of ethers.

4.5 Sulfides

4.5.1 Structure and Nomenclature of Sulfides

Sulfides, RSR', are sulfur analogs of ethers. They are named in the same way by following the same rules used for ethers, with sulfide used in place of ether for simple compounds and alkylthio used in place of alkoxy for more complex compounds (Figure 4.16).

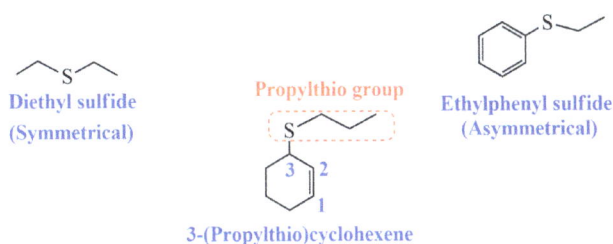

Diethyl sulfide
(Symmetrical)

Propylthio group

3-(Propylthio)cyclohexene

Ethylphenyl sulfide
(Asymmetrical)

Figure 4.16: Examples of sulfides.

4.5.2 Preparation and Reactions of Sulfides

Sulfides are prepared by treatment of primary or secondary alkyl halides with thiolate ion, RS⁻. The reaction occurs by an S_N2 mechanism, analogous to Williamson's synthesis of ethers. Thiolate anions are among the best nucleophiles known, and yields are usually high in these S_N2 reactions.

Dialkyl sulfides are good nucleophiles, and they react rapidly with primary alkyl halides by S_N2 mechanism to give trialkylsulfonium salts (R_3S^+).

Sulfides are also easily oxidized. For example, when methyl phenyl sulfide is treated with hydrogen peroxide, H_2O_2, at room temperature the corresponding methyl phenyl sulfoxide is formed, and further oxidation with peroxy acid yields a methyl phenyl sulfone, as shown in Figure 4.17.

4.6 Heterocyclic Compounds Containing Nitrogen, Oxygen, and Sulfur

Heterocyclic compounds have one or more hetero atoms such as nitrogen, oxygen, and sulfur in their structures. There are three to eight-membered structure heterocyclic compounds as summarized in Table 4.2. They can be classified (Figure 4.18) into aliphatic (cyclic amines, cyclic amides, cyclic ethers, and cyclic thioethers), selected examples are shown in Figure 4.19, and aromatic (benzene analogous), selected examples are presented in Figure 4.20.

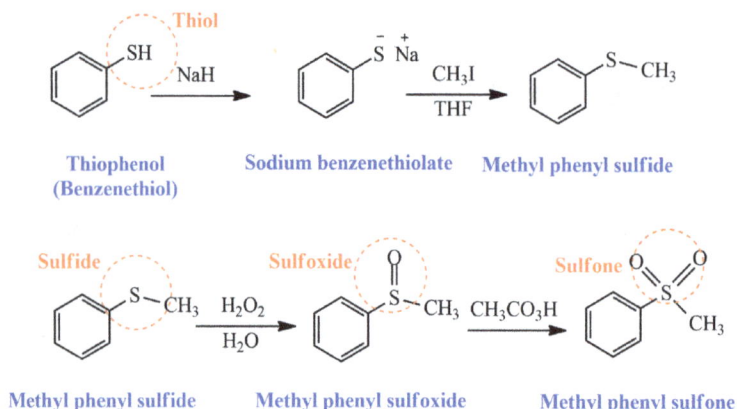

Figure 4.17: Preparation and reactions of sulfides.

Table 4.2: Selected examples of heterocyclic compounds.

Type	Example
Three-membered heterocycles	Aziridine and oxirane
Four-membered heterocycles	Oxetane
Five-membered heterocycles	Pyrazole, thiazole, isoxazole, furan, pyrrole, 1,3-dioxolane, oxadiazole, and tetrazole
Six-membered heterocycles	Pyridine, piperidine, and thiazine
Seven-membered heterocycles	Thiepine
Eight-membered heterocycles	Azocine

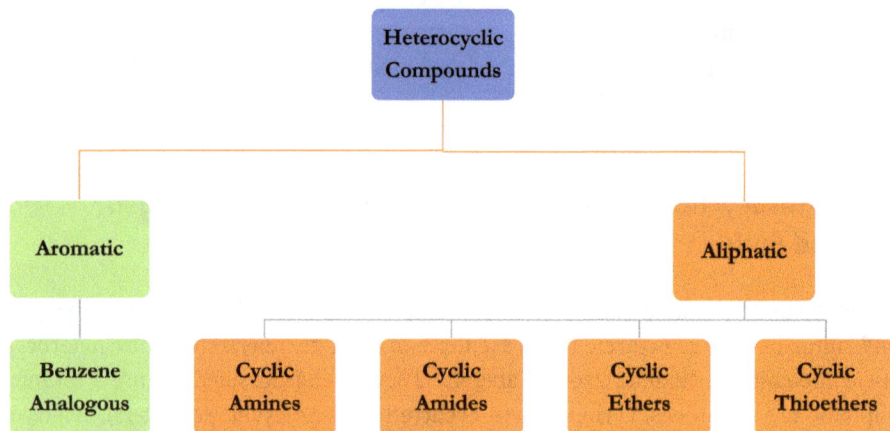

Figure 4.18: Classification of heterocyclic compounds.

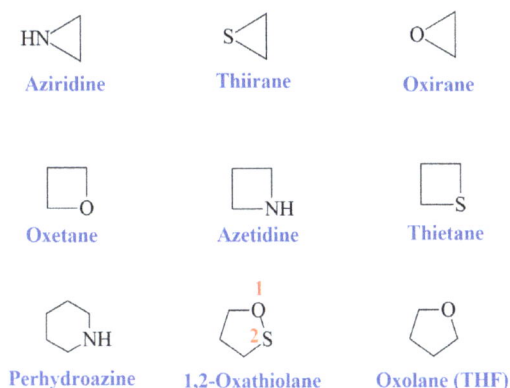

Figure 4.19: Selected examples of aliphatic heterocyclic compounds.

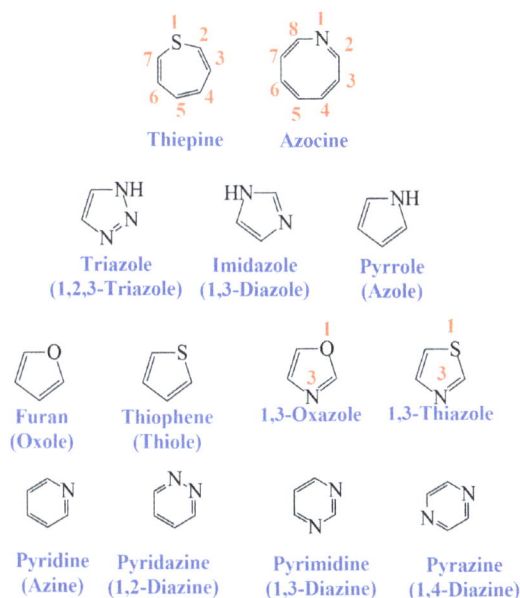

Figure 4.20: Selected examples of aromatic heterocyclic compounds.

Other examples of the heterocyclic compounds are the fused ring heterocycles such as quinoline, isoquinoline,, and indole that occur commonly in nature and many members of this class have pronounced biological activity (Figure 4.21).

In addition to the above-mentioned heterocyclic compounds, the bases of nucleic acids (nucleobases) are considered as one of the most important members of this family. They are the derivatives of purine and pyrimidine compounds. The purine once

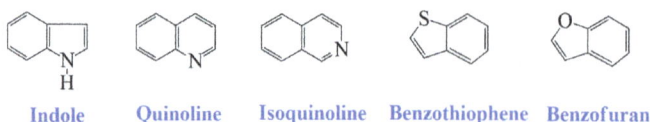

Indole   Quinoline   Isoquinoline   Benzothiophene   Benzofuran

Figure 4.21: Selected fused ring heterocyclic compounds.

are adenine (**A**) and guanine (**G**), and the derivatives of pyrimidine are thymine (**A**, 5-methyluracil), cytosine (**C**), and uracil (**U**) (Figure 4.22).

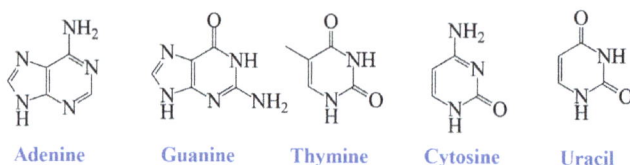

Adenine   Guanine   Thymine   Cytosine   Uracil

Figure 4.22: Nucleobases (nucleic acid bases) heterocyclic compounds.

4.6.1 Nomenclature of Heterocyclic Compounds

The original names of heterocyclic organic compounds are based on their preparation or their origin. For example, the heterocycle picoline was first prepared from coal tar and its name was derived from the Latin word pictus which means tarry. Heterocyclic compounds naming is also based on their source.

For example, the heterocyclic furfural was first isolated from barn oil, and its name was derived from that.

In general, to name heterocyclic compounds correctly three factors must be considered.

– Ring size
– Type of the heteroatom
– Position of the heteroatom

The heterocyclic compound name can also be derived from the carbocyclic name by introducing the prefix of the heteroatom oxygen (oxa), sulfur (thia), and nitrogen (aza). If more than one heteroatom of the same type exists in the structure, the numbering starts at the saturated one. If more than one type of heteroatoms exists in the structure the priority order shown in Figure 4.23 is followed and examples are shown in Figure 4.24.

Heterocyclic compounds can also be named by following Arthur Hantzsch and Oskar Widman's rules. For example, the 3–10-membered rings are named by combining the appropriate prefix (or prefixes) that indicates the type and position of the heteroatom present in the ring with the suffix that represents both the ring size and the degree

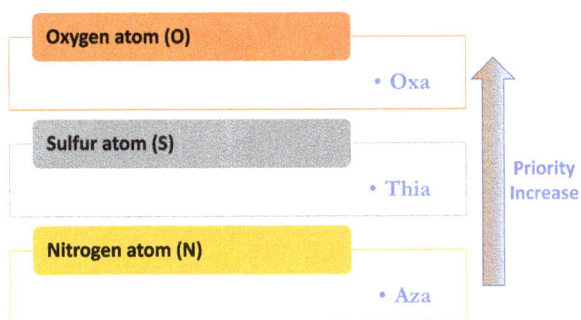

Figure 4.23: The priority order of selected heteroatoms.

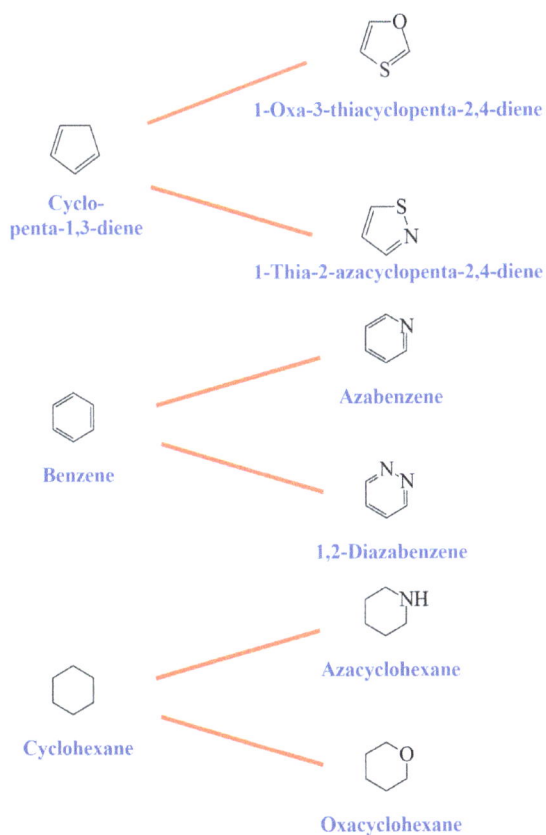

Figure 4.24: Examples of heterocyclic compounds compared with corresponding carbocyclic compounds.

of unsaturation. In addition, the suffixes distinguish between nitrogen-containing hetero-cycles and non-nitrogen-containing heterocycles as shown in Table 4.3. This can be summarized as follows:

IUPAC name = locants + prefix + suffix

To have the full name the following steps must be considered:
–   Identify and choose the corresponding prefix of the heteroatom as shown in Figure 4.23.
–   Numbering starts from the heteroatom position. So the heteroatom always gets position 1.
–   A multiplicative prefix (*di, tri*) and locants are used when two or more similar heteroatoms are contained in the ring (two nitrogen indicated by *diaza*) and the numbering starts at a saturated rather than an unsaturated atom.
–   Both prefixes are used if two heteroatoms are there in the structure and locants should be used to indicate the relative position of the heteroatoms.
–   If prefixes such as *oxa* and *aza* are combined the two vowels may come together so the vowel at the end of the first part should be omitted like *oxaza*.
–   The numbering should start at the heteroatom of the highest priority and at the same time give the smallest possible numbers to other heteroatoms in the ring.

Table 4.3: Suffixes for nitrogen-containing heterocycles and heterocycles that don't contain nitrogen.

Ring size	Heterocyclic compound with nitrogen		Heterocyclic compound with no nitrogen	
	Un-saturated	**Saturated**	**Unsaturated**	**Saturated**
3	irine	iridine	irene	irane
4	ete	etidine	ete	etane
5	ole	olidine	ole	olane
6	ine	e.g., perhydroazine	in	ane
7	epine	e.g., perhydroazepine	epin	epane
8	ocine	e.g.,perhydroazocine	ocin	ocane

To understand the above-mentioned table go through the following examples:

Example 1

This ring contains (N) so the prefix is *aza*.
    The ring is three-membered and fully saturated so the suffix is *iridine*.

By combining the prefix and suffix, two vowels ended up together (*azairidine*); therefore the vowel on the end of the first part should be dropped. This gives the correct name *Aziridine.*

Example 2

This ring contains(O) prefix 1 *oxa* and (N) prefix 2 *aza.*

Locants, since (O) is a higher priority than (N) so it is in position *1* by default and the (N) is therefore at position *2*, this gives the combined prefixes as *1,2,-oxaaza* (note that the a in oxa is not dropped).

It is a five-membered, fully unsaturated ring with (N) the suffix is *ole.*

By combining the prefixes and the suffix and dropping the appropriate vowels we get the correct name as *1,2-oxazole.*

Example 3

The ring is six-membered, fully saturated with N.

Prefix *perhydro* followed by the name of fully unsaturated six-membered ring with nitrogen *azine.*

Thus the full name is *perhydroazine.*

4.6.2 Reactions of Heterocyclic Compounds

Pyrrole is obtained commercially either directly from coal tar or by treatment of furan with ammonia over an alumina catalyst at 400 °C. Furan is synthesized by catalytic loss of carbon monoxide (decarbonylation) from furfural, which is itself prepared by acidic dehydration of the pentose sugar.

Thiophene is made in industry by cyclization-sulfurization of butane or butadiene with sulfur or H_2S at 600 °C and also in the lab from the reaction of sodium succinate with phosphorus trisulfide. On the other hand, pyrrolidine is prepared by the reaction of 1,4-butanediol with ammonia in the presence of nickel oxide and alumina, and tetrahydrofuran (THF) is prepared by the catalytic hydrogenation (H_2, Pd-C) of furan. Tetrahydrothiophene is prepared by reacting either furan or 1,4-dihydroxybutane with hydrogen sulfide in the presence of alumina at high temperatures. As for the six-membered hetero-cyclic compounds, pyridine is the most commonly known one. It is a stronger base than pyrrole and a common solvent in organic chemistry. Pyridine ring does not undergo elec-trophilic aromatic substitution reactions easily due to its low reactivity resulting from the electron-withdrawing inductive effect of the electronegative nitrogen.

For example, halogenation and sulfonation are carried out under drastic condi-tions. Nitration gives a very low yield, and Friedel-Crafts reactions are not successful at all. Reactions when occur usually give the three-substituted product. Selected het-erocycles' reactions are presented in Figure 4.25.

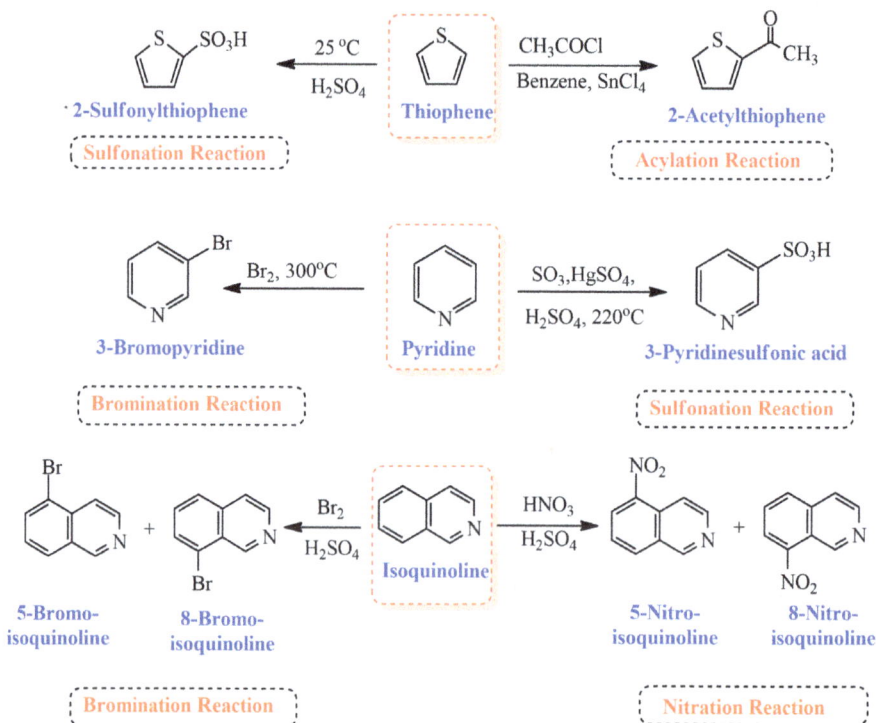

Figure 4.25: Selected heterocyclic compounds' reactions.

4.7 Essential Terms

Alcohols Compounds that have hydroxyl groups bonded to saturated sp^3-hybridized carbon atoms.

Aza One nitrogen replaces one carbon in the ring.

Carbocyclic rings Rings that have only carbon atoms.

Crown ethers A class of organic compounds that have large rings containing a number of ether linkages.

18-Crown-6-ether A cyclic ether that has 18 atoms and 6 of them are oxygen.

Dehydration of alcohols The removal of water from alcohol to form alkene.

Diaza Two nitrogens replace two carbons in the ring.

Five-membered saturated heterocycles Cyclic compounds that have five atoms in their ring and at least one hetero atom and only saturated bonds exist in the structure.

Five-membered unsaturated heterocycles Cyclic compounds that have five atoms in their ring and at least one hetero atom in addition to unsaturated bonds.

Fused-ring heterocycles Heterocyclic compounds containing two or more rings that share two atoms and one bond in common.

Heteroatoms Oxygen, sulfur, and nitrogen.

Heterocyclic rings Rings that have one or more different hetero atoms such as nitrogen, oxygen, and sulfur in addition to carbon.

Nucleobases Nucleic acids bases are found in DNA and RNA.

Oxa One oxygen replaces one carbon in the ring.

Oxidation of alcohols Converting primary alcohols to aldehydes or carboxylic acid and secondary alcohols to ketones with oxidizing agents.

PCC The mild oxidizing agent pyridinium chlorochromate.

PDC The mild oxidizing agent pyridinium dichromate.

Six-membered heterocycles Cyclic compounds that have six atoms in their ring and at least one hetero atom.

Sulfides (RSR') Sulfur analogs of ethers.

Symmetrical ether An ether that has two identical groups.

Unsymmetrical ether An ether that has two different groups.

Thia One nitrogen replaces one carbon in the ring.

Thiols (RSH) Sulfur analogs of alcohols.

Triaza Three nitrogens replace three carbons in the ring.

4.8 Problems

4.8.1   Briefly describe the general reaction for the chlorination of thiophene with chlorine (Cl_2).

4.8.2   Draw the structure of propanethiol and describe its oxidation with bromine (Br_2).

4.8.3 Classify the following amines into primary, secondary, and tertiary:

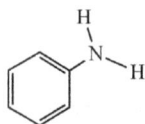

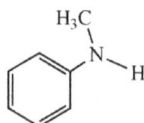

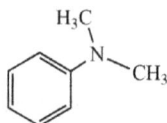

4.8.4 Explain why secondary amines can form hydrogen bonds among themselves.

4.8.5 Draw the chemical structures of the following thiols and sulfides:
i. Diethyl sulfide
ii. Ethyl methyl sulfide
iii. 2-Mercaptohexanol
iv. 1-Decanethiol

4.8.6 Complete the missing information in the following reaction:

$$RNH_2 \xrightarrow{RX} \ ? \ \xrightarrow{RX} \ ? \ \xrightarrow{RX} R_4\overset{+}{N}\ \overset{-}{X}$$

4.8.7 What kind of heteroatoms do organic heterocyclic compounds have?

4.8.8 Briefly describe the acylation of thiophene.

4.8.9 Name the following heterocyclic compounds:

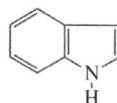

4.8.10 Which of the following amines and thiols are aromatic?

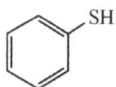

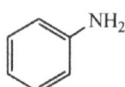

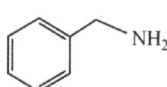

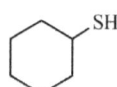

4.8.11 How are the sulfur analogues of ethers named?

4.8.12 How to name the following compound?

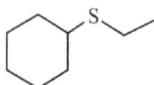

4.8.13 Draw the chemical structure of ethyl propyl sulfide.

4.8.14 Draw the chemical structure of 2-methylthiocyclopent-1,3-diene.

4.8.15 Draw the chemical structure of azine.

4.8.16 Draw the chemical structure of 1,2-oxazole.

4.8.17 In which ring positions do the electrophilic substitution take place in isoquinoline.

4.8.18 Draw the chemical structure of the alkylisothiourea salt.

4.8.19   What is the full name and the chemical structure of BME?

4.8.20   How are the sulfur analogues of alcohols named?

4.8.21   Give an example for symmetrical sulfides.

4.8.22   Give an example for cyclic ethers.

4.8.23   What does 18-Crown-6-ether stand for?

4.8.24   Draw the chemical structure for 18-crown-6-ether.

4.8.25   Draw the chemical structure of the major organic product of the following reaction:

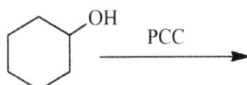

4.8.26   What will be the major organic product when 2-hexanol is treated with potassium dichromate dilute sulphuric acid?

4.8.27   Does 4,5-dimethylheptan-2-ol classified as primary or secondary or tertiary alcohol?

4.8.28   Give the IUPAC name for the following compound $(CH_3)_2CHCH_2OH$.

4.8.29   What is the major intermolecular attraction responsible for the relatively high boiling points of alcohols?

4.8.30   What is the name of the following compound?

****

Chapter 5
Aldehydes and Ketones

Objectives

After studying this chapter, learners will be able to:
- Write the common and IUPAC names of aldehydes and ketones.
- Draw the chemical structures of the aldehydes and ketones.
- Describe the important methods of preparation and reactions of these classes of compounds.
- Correlate physical properties of aldehydes and ketones with their structures.
- Explain mechanisms of selected reactions of aldehydes and ketones.

5.1 Structure and Nomenclature of Aldehydes and Ketones

Aldehydes can be named by exchanging -**al** for the -**e** of the corresponding alkane. Numbering must always start from the aldehyde carbon.

Ketones are also named by exchanging -one for the -e of the corresponding alkane. Numbering must start in the direction that gives the ketone carbon the lowest possible number. Selected examples of aldehydes and ketones with 3–10 carbon atoms in their structure, along with their IUPAC names, are listed in Table 5.1. Other examples of aldehydes and ketones with both common and IUPAC names are presented in Figure 5.1.

5.2 Physical Properties of Aldehydes and Ketones

Low-molecular weight aldehydes and ketones are either gases or liquids at room temperature. Their boiling points are higher than hydrocarbons and ethers of comparable molecular masses due to their weak molecular association arising from the dipole-dipole interactions. Additionally, their boiling points are lower than alcohols of similar molecular masses due to the absence of intermolecular hydrogen bonding.

The low-molecular weight aldehydes and ketones such as methanal, ethanal, and propanone are miscible with water in all proportions due to hydrogen bonding with water, but their solubility decreases by increasing the number of carbons in the chain. All aldehydes and ketones are fairly soluble in organic solvents like benzene, ether, methanol, and chloroform.

https://doi.org/10.1515/9783111382753-005

Table 5.1: Structures of aldehydes and ketones with 3–10 carbons.

Number of carbons Aldehyde/ketone	Aldehyde structure	Ketone structure
3 Propanal/2-propanone		
4 Butanal/2-butanone		
5 Pentanal/2-pentanone		
6 Hexanal/2-hexanaone		
7 Heptanal/2-hepatanone		
8 Octanal/2-octanone		
9 Nonanal/2-nonanone		
10 Decanal/2-decanone		

5.3 Preparation of Aldehydes and Ketones

There are many reported methods for the preparation of aldehydes and ketones. The selected ones are listed in Table 5.2.

Benzene carbaldehyde
OR
Benzaldehyde

Methanal
OR
Formaldehyde

Ethanal
OR
Acetaldehyde

Propanal
OR
Propionaldehyde

Cyclohexanecarbaldehyde
OR
Cyclo hexanal

Butanone
OR
Ethyl methyl
ketone

Propanone
OR
Acetone
OR
Dimethyl ketone

1-phenylethanone
OR
Acetophenone
OR
Methyl phenyl ketone

Diphenylmethanone
OR
Benzophenone
OR
Diphenyl ketone

Benzene carbaldehyde
OR
Benzaldehyde

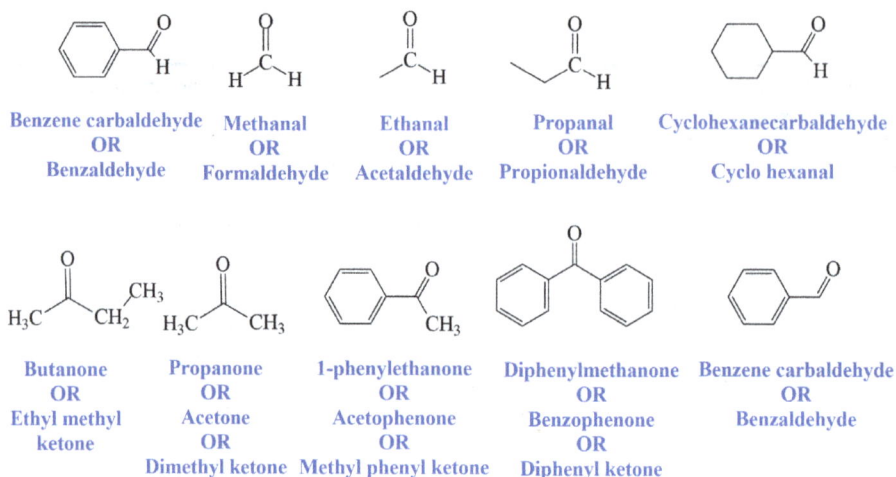

Figure 5.1: Selected examples of aldehydes and ketones.

Table 5.2: Selected methods for the preparation of aldehydes and ketones.

Preparation	Aldehyde	Ketone
From alcohols "Oxidation"	Pyridinium dichromate (PDC) or pyridinium chlorochromate (PCC) in anhydrous media such as dichloromethane oxidizes primary alcohols to aldehydes while avoiding over-oxidation (Jones oxidation) to carboxylic acids.	Many oxidizing agents are available for converting secondary alcohols to ketones. PDC or PCC (for sensitive alcohols) as well as Cr (VI)-based agents such sodium or potassium dichromate in acetic acid or chromium trioxide in sulfuric acid can also be used.
Swern oxidation	Primary and secondary alcohols can be oxidized to aldehydes and ketones, respectively, by using oxalyl chloride, dimethyl sulfoxide, and triethylamine mixture.	
From alkenes "ozonolysis"	Ozonolysis of alkenes followed by reaction with zinc dust and water gives aldehydes.	
From alkynes "Hydration"	Addition of water to ethyne in the presence of H_2SO_4 and $HgSO_4$ gives acetaldehyde. All other alkynes give ketones. Reaction occurs through an enol intermediate formed by Markovnikov's addition of water to multiple bonds.	
From acyl chlorides "Hydrogenation" "Grignard reagent"	Acyl chlorides (acid chloride) are hydrogenated over catalyst, palladium on barium sulfate to give aldehydes, Rosenmund reduction.	Treatment of acyl chlorides with dialkylcadmium, prepared by the reaction of cadmium chloride with Grignard reagent, gives ketones.

Table 5.2 (continued)

Preparation	Aldehyde	Ketone
From nitriles "Reduction" "Nucleophilic acyl substitution" "Nucleophilic addition"	Nitriles are reduced to corresponding imine with stannous chloride in the presence of hydrochloric acid, which on hydrolysis give corresponding aldehyde, Stephen reaction.	Treating a nitrile with Grignard reagent followed by hydrolysis yields a ketone.
From esters "Reduction"	Esters are also reduced to aldehydes with diisobutyl-aluminum hydride (DIBAL-H).	
From methylbenzene "Oxidation"	Chromyl chloride (CrO_2Cl_2) oxidizes methyl group of toluene to a chromium complex, which on hydrolysis gives corresponding benzaldehyde, Etard reaction.	
Gattermann-Koch reaction	Benzenes and their derivatives are converted to benzaldehydes or substituted benzaldehydes by treatment with carbon monoxide and hydrogen chloride in the presence of anhydrous aluminum chloride or cuprous chloride.	
From aromatic compounds "Friedel-Crafts Acylation"		Acyl chlorides and carboxylic acid anhydrides acylate aromatic rings in the presence of aluminum chloride. The reaction is an electrophilic aromatic substitution where the acylium ion is generated to attack the ring and form the ketone.

5.4 Reactions of Aldehydes and Ketones

The carbonyl group, C=O, is a very important functional group in organic chemistry. It exists in nature in many organic compounds and also in many pronounced pharmaceutical agents and many of the synthetic chemicals that touch our everyday lives.

The high electronegativity of oxygen relative to carbon results in a polarized C=O double bond. The carbonyl group (C=O) polarization has a great effect on the chemical reactivity of the carbonyl compounds such as aldehydes and ketones.

Due to that reactivity, carbonyl oxygen reacts with acids and electrophiles, while carbonyl carbon reacts with bases and nucleophiles. Table 5.3 summarizes selected aldehydes and ketones reactions.

Table 5.3: Selected reactions of aldehydes and ketones.

Addition of Hydrogen Cyanide (HCN)

Aldehydes and ketones react with hydrogen cyanide (HCN) to yield cyanohydrins.

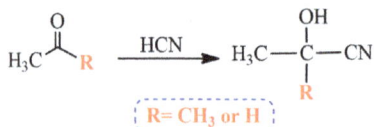

$$H_3C \overset{O}{\underset{}{\overset{\|}{C}}} R \xrightarrow{\text{HCN}} H_3C-\overset{OH}{\underset{R}{\overset{|}{C}}}-CN$$

R= CH₃ or H

Addition of Grignard Reagent (RMgX)

Nucleophilic addition of R- to C=O group produces a tetrahedral magnesium alkoxide intermediate which upon protonation by dilute acid such as H_2SO_4 or HCl gives neutral alcohol.

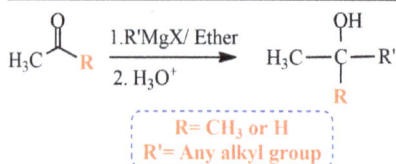

$$H_3C \overset{O}{\underset{}{\overset{\|}{C}}} R \xrightarrow[\text{2. } H_3O^+]{\text{1. R'MgX/ Ether}} H_3C-\overset{OH}{\underset{R}{\overset{|}{C}}}-R'$$

R= CH₃ or H
R'= Any alkyl group

Addition of Ammonia (NH₃)

Ammonia adds to the carbonyl group of aldehydes and ketones to form imines.

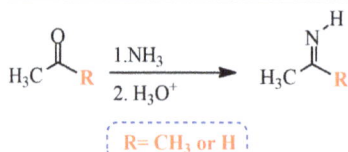

$$H_3C \overset{O}{\underset{}{\overset{\|}{C}}} R \xrightarrow[\text{2. } H_3O^+]{\text{1. NH}_3} H_3C \overset{N^{\diagup H}}{\underset{}{\overset{\|}{C}}} R$$

R= CH₃ or H

Reduction to Alcohols

Aldehydes and ketones are reduced in ether or THF to primary and secondary alcohols, respectively, by sodium borohydride or lithium aluminum hydride or by catalytic hydrogenation.

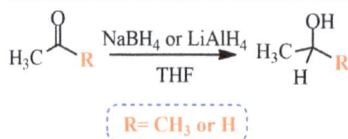

$$H_3C \overset{O}{\underset{}{\overset{\|}{C}}} R \xrightarrow[\text{THF}]{\text{NaBH}_4 \text{ or LiAlH}_4} H_3C \overset{OH}{\underset{H}{\overset{|}{C}}} R$$

R= CH₃ or H

Reduction to Hydrocarbons

Aldehydes and ketones are reduced to hydrocarbons (C=O to CH₂) by treatment with zinc amalgam and concentrated hydrochloric acid (*Clemmensen Reduction*) or with hydrazine followed by heating with sodium or potassium hydroxide in ethylene glycol (*Wolff-Kishner Reduction*).

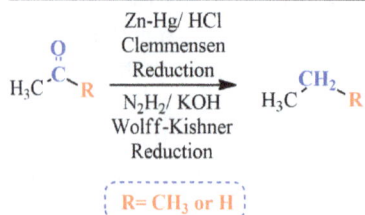

$$H_3C \overset{O}{\underset{}{\overset{\|}{C}}} R \xrightarrow[\substack{N_2H_2/ \text{ KOH} \\ \text{Wolff-Kishner} \\ \text{Reduction}}]{\substack{\text{Zn-Hg/ HCl} \\ \text{Clemmensen} \\ \text{Reduction}}} H_3C \overset{CH_2}{\diagdown} R$$

R= CH₃ or H

Table 5.3 (continued)

Oxidation to Carboxylic Acids

Aldehydes are easily oxidized to carboxylic acids on treatment with common oxidizing agents such as potassium permanganate ($KMnO_4$), potassium dichromate ($K_2Cr_2O_7$), or chromium trioxide in sulfuric acid (CrO_3-H_2SO_4) known as Jones oxidation. Ketones are generally oxidized under vigorous conditions with strong oxidizing agents and at elevated temperatures. Their oxidation involves carbon-carbon bond cleavage to afford a mixture of carboxylic acids having lesser number of carbon atoms than the parent ketone.

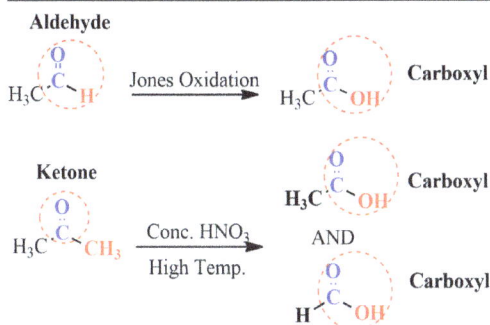

Aldol Condensation

Aldehydes and ketones having at least one α-hydrogen undergo a reaction in the presence of dilute alkali as catalyst to form β-hydroxy aldehydes (**aldol**) or β-hydroxy ketones (**ketol**), respectively. The aldol and ketol readily lose water to give α,β-unsaturated carbonyl compounds which are aldol condensation products, and the reaction is called aldol condensation.

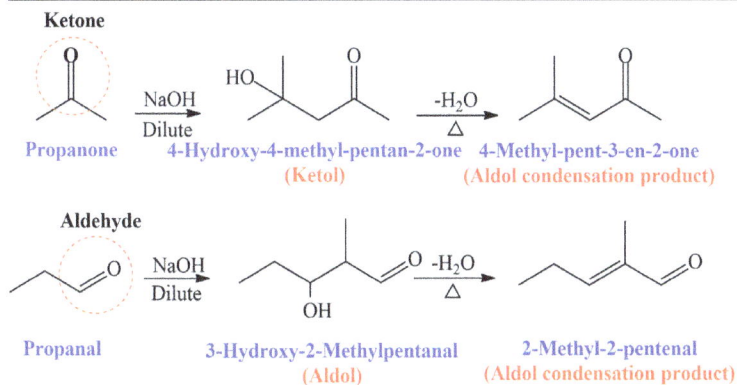

Table 5.3 (continued)

Cannizzaro Reaction

Aldehydes, which do not have an α-hydrogen atom, undergo self-oxidation and reduction reaction on heating with concentrated alkali. One molecule of the aldehyde is reduced to alcohol while another one is oxidized to carboxylic acid salt. Good examples are benzaldehyde, methanal, and 2,2-dimethylpropanal.

5.5 Essential Terms

Acetaldehyde Ethanal.

Acetophenone Methyl phenyl ketone.

Aldehydes Carbonyl compounds with the general formula R-CHO.

Aldol reaction Formation of an aldol from an aldehyde with α-hydrogen in the presence of dilute alkali.

Benzophenone Diphenyl ketone.

Cannizzaro reaction The redox reaction of an aldehyde to carboxylic acid and alcohol in the presence of sodium hydroxide.

Clemmensen reduction Reduction of aldehydes and ketones to hydrocarbons by treatment with zinc amalgam and concentrated hydrochloric acid.

Etard reaction The oxidation of the toluene methyl group by chromyl chloride (CrO_2Cl_2) to give a chromium complex, which on hydrolysis gives corresponding benzaldehyde.

Gattermann-Koch reaction The transformation of benzenes and their derivatives to benzaldehydes or substituted benzaldehydes by treatment with carbon monoxide and hydrogen chloride in the presence of anhydrous aluminum chloride or cuprous chloride.

Grignard reagent An organomagnesium halide with a general structure "RMgX."

Hydration of alkynes to ketones Formation of ketones from alkynes by addition of water.

Jones oxidation Over-oxidation of alcohols to carboxylic acids by chromium trioxide in sulfuric acid (CrO_3-H_2SO_4-H_2O).

Ketones Carbonyl compounds with the general formula R-CO-R.

PCC Pyridinium chlorochromate.

PDC Pyridinium dichromate.

Rosenmund reduction The hydrogenation of acyl chlorides (acid chlorides) over palladium on barium sulfate to give aldehydes.

Stephen reaction The reduction of nitriles to corresponding imines with stannous chloride in the presence of hydrochloric acid, followed by hydrolysis to give corresponding aldehyde.

Swern oxidation The use of dimethyl sulfoxide as the oxidizing agent to convert primary and secondary alcohols to aldehydes and ketones.

Wolff-Kishner reduction Reduction of aldehydes and ketones to hydrocarbons by treatment with hydrazine followed by heating with sodium or potassium hydroxide in ethylene glycol.

5.6 Problems

5.6.1   Give the IUPAC name for the following compounds:
   i. $C_2H_5COC_2H_5$
   ii. $C_2H_5COCH_3$
   iii. $CH_3(CH_2)_4CHO$
   iv. CH_3CH_2CHO

5.6.2   Draw the skeletal formulas of the following compounds:
   i.   4-Chloro-2-pentanone
   ii.   3-Methyl pentanal
   iii.   Cyclohexanone
   iv.   3-Methyl-3-butenal
   v.   3-Heptanone
   vi.   2,4,6-Trinitrobenzaldehyde

5.6.3   Predict the major product(s) when each of the following alcohols reacts with $Na_2Cr_2O_7$, H_2SO_4:
   i.   Hexan-1-ol
   ii.   Octan-3-ol
   iii.   4-Hydroxynonane

5.6.4   Which of the following compounds would undergo the Cannizzaro reaction?
   i.   Benzaldehyde
   ii.   Ethanal
   iii.   2-Methylpentanal

5.6.5   What would be the final product from the reaction of benzaldehyde with zinc amalgam and concentrated hydrochloric acid?

5.6.6   Name the following compounds according to IUPAC nomenclature system:
   i. $CH_3CH_2CH=CHCHO$
   ii. $CH_3COCH_2CH_2CH_3$

5.6.7   Draw the chemical structure of 4,4′-dihydroxybenzophenone.

5.6.8   Show the acid-base complex between MgX and the carbonyl oxygen.

5.6.9   What do aldol and ketol terms stand for?

5.6.10   Which part of the carbonyl group reacts with electrophiles and acids?

5.6.11   Do α-Hydroxyketones give a positive Tollen's test? Explain why.

5.6.12   Complete the following reactions:

5.6.13   Suggest a simple chemical test to distinguish between butanal and butanone.

5.6.14   Draw the structure of 2,4-heptanedione.

5.6.15   Draw the chemical structure of oxetane-2-carbaldehyde.

5.6.16   What are the possible names for the following compound?

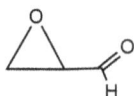

5.6.17   Is propanal liquid or solid at room temperature?

5.6.18   What is Jones reagent?

5.6.19   What is the final product of the oxidation of 2-pentanol with PCC?

5.6.20   What is the final product of the oxidation of primary alcohols with Swern reagent?

5.6.21   What kind of reagents can be used to carry out the following transformations?
    i.   Hexan-1-ol to hexanal
    ii.   Cyclopentanol to cyclopentanone
    iii. p-Chlorotoluene to p-chlorobenzaldehyde
    iv.   Ethanenitrile to ethanal

5.6.22   What is the aldol condensation product of butanal?

5.6.23   Show how 2-methyl-2-pentenal is prepared from propanal.

5.6.24   Describe the Cannizzaro reaction and what kind of an aldehyde is required to achieve it?

5.6.25   What is the difference between aldol condensation and cross aldol condensation?

5.6.26   Give an example for ketols.

5.6.27   Can tertiary alcohols undergo Swern oxidation?

5.6.28   Briefly describe the Gattermann-Koch reaction.

5.6.29   Describe the preparation of cyanohydrins from aldehydes.

5.6.30   What is Jones oxidation?

****

Chapter 6
Carboxylic Acids and Their Derivatives

Objectives

After studying this chapter, learners will be able to:
- Write the common and IUPAC names of carboxylic acids and their derivatives.
- Draw the chemical structures of the compounds containing carboxyl groups such as long-chain carboxylic acids and amino carboxylic acids.
- Describe the important methods of preparation and reactions of esters, amides, acid chlorides, and acid anhydrides.
- Correlate physical properties and chemical reactions of carboxylic acids, with their structures.

6.1 Carboxylic Acids

Carboxylic acids have two functional groups, namely the carbonyl and the hydroxyl. The carboxyl group (–COOH) is the parent group of a family of compounds called acyl compounds or carbonyl compounds or carboxylic acid derivatives.

6.1.1 Structure and Nomenclature of Carboxylic Acids

The COOH group is the functional group of carboxylic acids. It is simply made up of a carbonyl group (C=O) and a hydroxyl group (OH). It can be presented in three alternative structures as shown in Figure 6.1.

-COOH   -CO$_2$H   **Figure 6.1:** Carboxyl group.

In IUPAC nomenclature, the name (*in blue color*) of a carboxylic acid is derived by exchanging the -oic acid for the -e of the corresponding alkane, and numbering must start at the carboxyl carbon as shown in Figure 6.2.

Not only IUPAC names of carboxylic acids are used but also common names for many carboxylic acids are still in use. Selected examples are listed in Table 6.1.

https://doi.org/10.1515/9783111382753-006

Butanoic acid
(Butyric Acid)

Pentanoic acid
(Valeric Acid)

Heptanoic acid
(Enanthic acid)

2-Propenoic acid
(Acrylic acid)

2-Hydroxypropanoic acid
(Lactic acid)

2-Butenoic acid
(Crotonic acid)

2-Hydroxy-2-phenyl-ethanoic acid
(Mandelic acid)

2,3-Dimethylpentanoic acid
(α,β-Dimethylvaleric acid)

Hexanoic acid
(Caproic Acid)

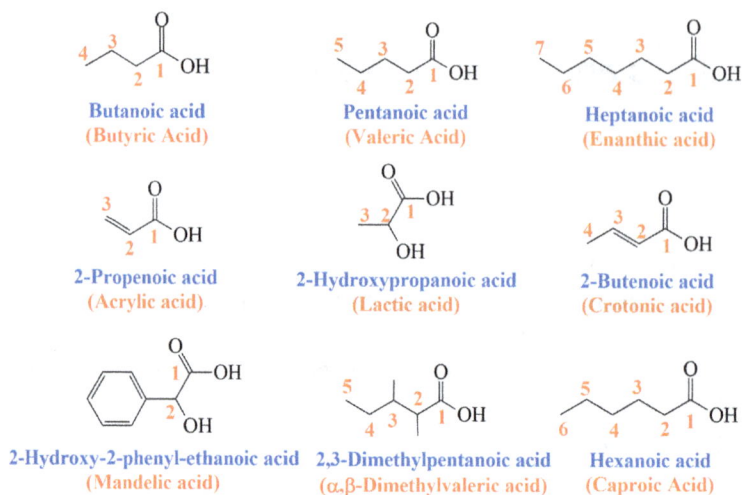

Figure 6.2: Selected examples of carboxylic acids.

Table 6.1: IUPAC and common names for selected carboxylic acids.

IUPAC name	Chemical structure	Common name
Methanoic acid		Formic acid
Ethanoic acid		Acetic acid
Propanoic acid		Propionic acid
Butanoic acid		Butyric acid
Pentanoic acid		Valeric acid

Table 6.1 (continued)

IUPAC name	Chemical structure	Common name
Hexanoic acid		Caproic acid
Heptanoic acid		Enanthic acid
Octanoic acid		Caprylic acid
Nonanoic acid		Pelargonic acid
Decanoic acid		Capric acid
2-Hydroxybenzoic acid		Salicylic acid
Benzenecarboxylic acid		Benzoic acid

6.1.2 Physical Properties of Carboxylic Acids

Carboxylic acids have higher boiling points than those of hydrocarbons and oxygen-containing compounds of comparable size. Carboxylic acids with 1–4 carbon atoms are miscible with water in all proportions because the hydrophilic character of the carboxyl group more than counterbalances the hydrophobic character of the hydrocarbon chain. As the size of the hydrocarbon chain increases relative to the size of the carboxyl group, water solubility decreases. For example, while the solubility of acetic acid is infinite, the solubility of hexanoic acid in water drops to 1.0 g/100 g water and that of decanoic acid is only 0.2 g/100 g water. Carboxylic acids also interact with water molecules by hydrogen bonding through both their carbonyl and hydroxyl groups. They can also form strong hydrogen bonds with each other and exist as dimers as shown in Figure 6.3. Carboxylic acids (pK_a 4–5) are stronger acids than alcohols (pK_a 16–18). Substitution at the α-carbon of an atom or a group of atoms of higher electronegativity than carbon further increases the acidity of carboxylic acids.

A single chlorine substituent increases acid strength by nearly 100 times. Trichloroacetic acid (pK_a 0.66) is a stronger acid than H_3PO_4 (pK_a 2.16).

The liquid carboxylic acids, from propanoic acid to decanoic acid, have extremely foul odors, about as bad as those of thiols.

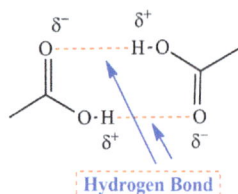

Figure 6.3: Hydrogen bond formation in the acetic acid dimer.

6.1.3 Long-Chain Carboxylic Acids (Fatty Acids)

The size of carboxylic acid chains ranges from C12 to C26 for long-chain carboxylic acids, C8 and C10 carbon atoms for medium-chain carboxylic acids, and C4 and C6 for short-chain carboxylic acids. Long-chain carboxylic acids are known as fatty acids. They do not dissolve in water but dissolve in fat solvents such as hexane, ether, acetone, and chloroform and can be categorized into:

- Saturated long-chain carboxylic acids (Figure 6.4A) are the ones that contain only C–C single bonds. They are closely packed and have strong attractions between chains. They also have high melting points and exist as solids at room temperature.
- Unsaturated carboxylic acids (Figure 6.4B) are the second type of long-chain carboxylic acids. They have one or more C=C bonds and C-C bonds. Since they have few interactions between chains, they are not closely packed. They also have low melting points and exist as liquids at room temperature.

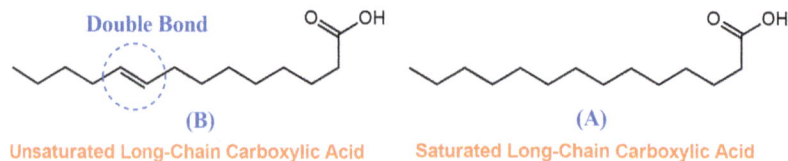

Figure 6.4: Saturated (A) and unsaturated (B) long-chain carboxylic acids.

Selected examples of common and well-known long-chain carboxylic acids are listed in Table 6.2.

Table 6.2: Selected examples of common and well-known long-chain carboxylic acids.

Common name	IUPAC name	Number of carbon atoms	Number of double bonds
Saturated carboxylic acids			
Lauric acid	Dodecanoic acid	12	0
Myristic acid	Tetradecanoic acid	14	0
Palmitic acid	Hexadecenoic acid	16	0
Stearic acid	Octadecanoic acid	18	0
Arachidic acid	Eicosanoic acid	20	0
Unsaturated carboxylic acids			
Palmitoleic acid	Hexadec-9-enoic acid	16	1
Oleic acid	Octadec-9-enoic acid	18	1
Linoleic acid	Octadeca-9,12-dienoic acid	18	2
Linolenic acid	Octadeca-9,12,15-trienoic acid	18	3
Arachidonic acid	Icosa-5,8,11,14-tetraenopic acid	20	4

6.1.4 Amino Carboxylic Acids (Amino Acids)

Carboxylic acids that have amino groups in their structure are known as amino carboxylic acids. They can be classified into 2-amino, 3-amino, 4-amino, and even 5-aminocarboxylic acids as shown in Figure 6.5.

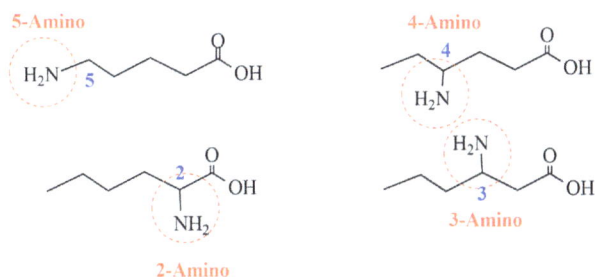

Figure 6.5: Classes of amino-carboxylic acids.

Among those, the 2-aminocarboxylic acids also known as amino acids are the most important ones. They are distinguished from each other by the type of side chain "R" present in their structure. Selected examples are listed in Table 6.3.

Table 6.3: Selected examples of 2-aminocarboxylic acids.

IUPAC name	Common name	Formula
2-Aminopropanoic acid	Alanine	$CH_3CH(NH_2)COOH$
2-Amino-5-guanidinopentanoic acid	Arginine	$NH_2C(=NH)NH[CH_2]_3CH(NH_2)-COOH$
2-Amino-3-carbamoylpropanoic acid	Asparagine	$NH_2COCH_2CH(NH_2)COOH$
2-Aminobutanedioic acid	Aspartic acid	$HOOCCH_2CH(NH_2)COOH$
2-Amino-3-mercaptopropanoic acid	Cysteine	$HSCH_2CH(NH_2)COOH$
2-Aminopentanedioic acid	Glutamic acid	$HOOCCH_2CH_2CH(NH_2)COOH$
2-Aminoethanoic acid	Glycine	$CH_2(NH_2)COOH$
2-Amino-3-methylpentanoic acid	Isoleucine	$CH_3CH_2CH(CH_3)CH(NH_2)COOH$
2-Amino-4-methylpentanoic acid	Leucine	$(CH_3)_2CHCH_2CH(NH_2)COOH$
2,6-Diaminhexanoic acid	Lysine	$NH_2[CH_2]_4CH(NH_2)COOH$
2-Amino-3-phenylpropanoic acid	Phenyl-alanine	$C_6H_5CH_2CH(NH_2)COOH$
2-Amino-3-hydroxypropanoic acid	Serine	$HOCH_2CH(NH_2)COOH$
2-Amino-3-hydroxybutanoic acid	Threonine	$CH_3CH(OH)CH(NH_2)COOH$
2-Amino-3-(4-hydroxyphenyl)propan-oic acid	Tyrosine	$HO-C_6H_5-CH_2CH(NH_2)COOH$
2-Amino-3-methylbutanoic acid	Valine	$(CH_3)_2CHCH(NH_2)COOH$

6.1.5 Reactions of Carboxylic Acids

Carboxylic acids are reduced to alcohols by $LiAlH_4$. They react with thionyl chloride to give acid chlorides and with alcohols in the presence of acid catalysts to form esters. They also react with ammonia at elevated temperatures to form amides and with acid chlorides in pyridine to produce acid anhydrides. All carboxylic acids, whether soluble or insoluble in water, react with NaOH, KOH, ammonia, and amines to form water-soluble salts. Examples of the above-mentioned reactions are presented in Figure 6.6.

6.2 Dicarboxylic Acids

Compounds that have two carboxyl groups in their structure are known as dicarboxylic acids. The two carboxyl groups can either be closer together or separated by a number of carbon atoms. According to the IUPAC naming system, aliphatic dicarboxylic acids are simply named as alkanedioic acids and the aromatic ones as benzenedicarboxylic acids. Common names are also in use for simple dicarboxylic acids. Chemical structures of selected examples of dicarboxylic acids are presented in Figure 6.7, and their IUPAC names, common names, and uses are listed in Table 6.4.

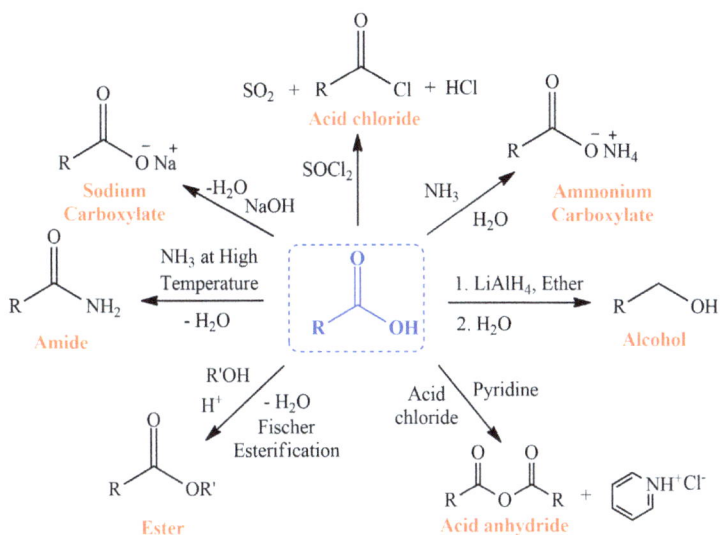

Figure 6.6: Reactions of carboxylic acids.

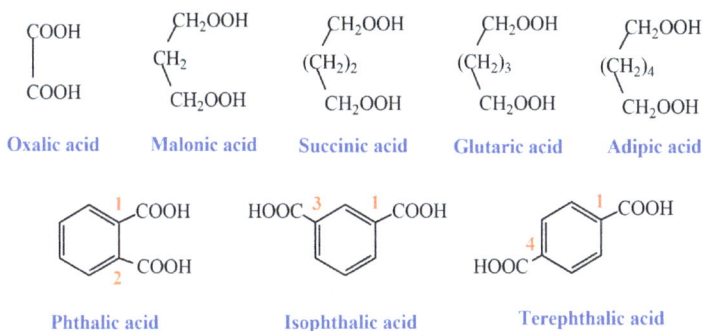

Figure 6.7: Selected examples of dicarboxylic acids.

Table 6.4: Selected examples of dicarboxylic acids.

IUPAC name	Common name	Uses
1,2-Benzenedicarboxylic acid	Phthalic acid	Used as plasticizer
1,3-Benzenedicarboxylic acid	Isophthalic acid	Used in making copolyester resins
1,4-Benzenedicarboxylic acid	Terephthalic acid	Used as a raw material to make terephthalate plasticizer
Ethanedioic acid	Oxalic acid	Used as bleaching agent
Propanedioic acid	Malonic acid	Used for making barbiturates
Butanedioic acid	Succinic acid	Used as precursor for making lacquers and dyes
Pentanedioic acid	Glutaric acid	Used in the production of polyols
Hexanedioic acid	Adipic acid	Used in the production nylon and urethane foams

6.3 Carboxylic Acid Derivatives

The carboxylic acid derivatives are classes of organic compounds derived from carboxylic acids and each has an acyl group (*RCO*- or *ArCO*-) attached to a halogen, oxygen, or nitrogen atom. The four classes are:

–   Carboxylic esters (RCO-OR or RCO-OAr).
–   Acid anhydrides (RCO-O-COR or ArCO-O-COAr).
–   Acid halides (acyl halides) mainly acid chlorides (RCO-Cl or ArCOCl).
–   Amides: primary (RCO-NH$_2$), secondary (RCO-NHR'), tertiary (RCO-NR'R"), and their aromatic derivatives.
–   Nitriles (R-C≡N).

6.3.1 Structure and Nomenclature of Carboxylic Acids Derivatives

Esters' names are derived from the names of the corresponding carboxylic acids and alcohols. The alcohol portion is named first and has the ending *–yl* and the carboxylic acid portion follows, and its name ends with *-ate* or *–oate*. Cyclic esters are called *lactones*. The IUPAC name of a lactone is formed by dropping the suffix *-oic* acid from the name of the parent carboxylic acid and adding the suffix *-olactone*. The location of the oxygen atom in the ring is indicated by a number if the IUPAC name of the acid is used and by a Greek letter α, β, γ, δ, ε, and so forth if the common name of the acid is used. Acid anhydrides are named by dropping the word acid from the carboxylic acid name and adding the word *anhydride.* Acid chlorides are named by dropping the *-ic acid* from the name of the carboxylic acid and adding *-yl chloride* or adding *carbonyl chloride* if derived from cycloalkane carboxylic acid.

Primary amides are named by replacing *-ic acid* in the carboxylic acid name with *amide*. Groups on the nitrogen are named as substituents and are given the locants *N*- or *N, N*-. Cyclic amides are given a special name *lactam*. Their common names are derived in a manner similar to those of lactones, with the difference that the suffix *-olactone* is replaced by *-olactam.*

Nitriles have cyano-group (C≡N) bonded to a carbon atom. Their IUPAC names follow the pattern *alkanenitrile* (e.g., *ethanenitrile*) and their common names are derived by dropping the suffix *-ic* or *-oic* acid from the name of the parent carboxylic acid and adding the suffix *-onitrile*. Selected examples of the above-mentioned derivatives are shown in Figure 6.8.

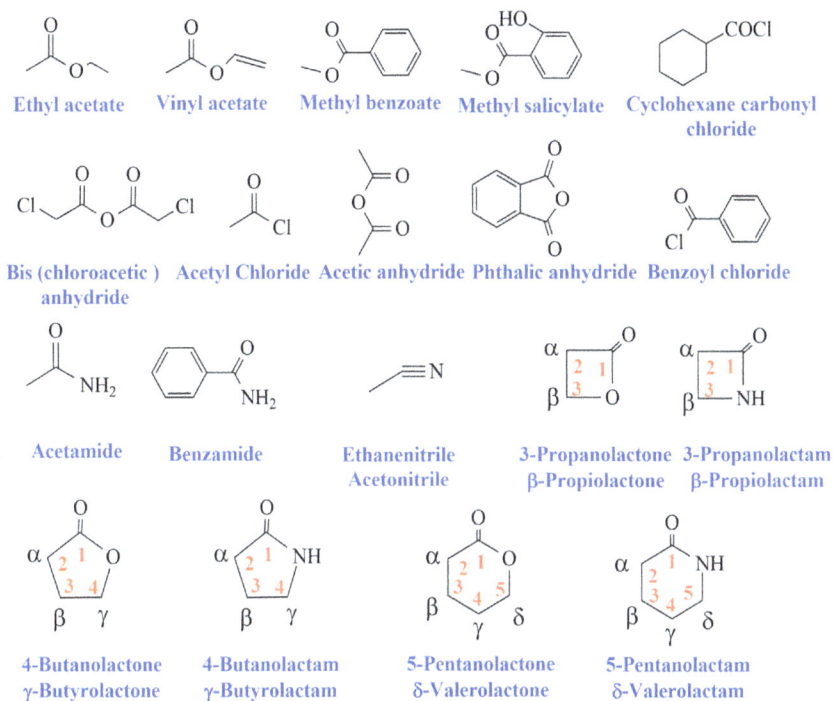

Figure 6.8: Selected examples of carboxylic acids derivatives.

6.3.2 Physical Properties of Carboxylic Acids Derivatives

Esters have lower boiling points than carboxylic acids because they cannot hydrogen bond to each other, but they can hydrogen bond to water and have appreciable water solubility.

Lower aliphatic anhydrides are colorless liquids, but higher aliphatic and aromatic anhydrides are colorless solids. They dissolve in organic solvents such as alcohols and ethers.

Acids chlorides are soluble in organic solvents and have lower boiling points than the corresponding acids.

Primary and secondary amides form very strong hydrogen bonds and have high melting and boiling points. Tertiary amides cannot form hydrogen bonds to each other and have lower melting and boiling points.

Nitriles, with low molecular weight, are liquids at room temperature. Low-molecular weight nitriles are soluble in water, and the solubility then decreases as chain length increases.

6.3.3 Reactions of Carboxylic Acids Derivatives

6.3.3.1 Reactions of Esters

Esters undergo hydrolysis when heated with water in the presence of strong acids or bases (saponification). The mechanism of hydrolysis is the reverse of Fischer esterification. Esters are readily reduced by lithium aluminum hydride to give two alcohols as shown in Figure 6.9.

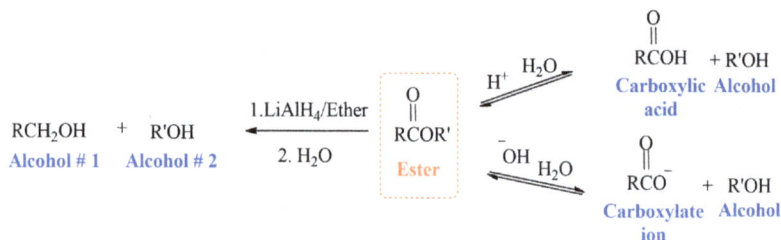

Figure 6.9: Reactions of esters.

6.3.3.2 Reactions of Carboxylic Acid Anhydrides

Acid anhydrides react with water to form carboxylic acids, with alcohols to form esters, and with amines to form amides. They can also be reduced to alcohols (Figure 6.10).

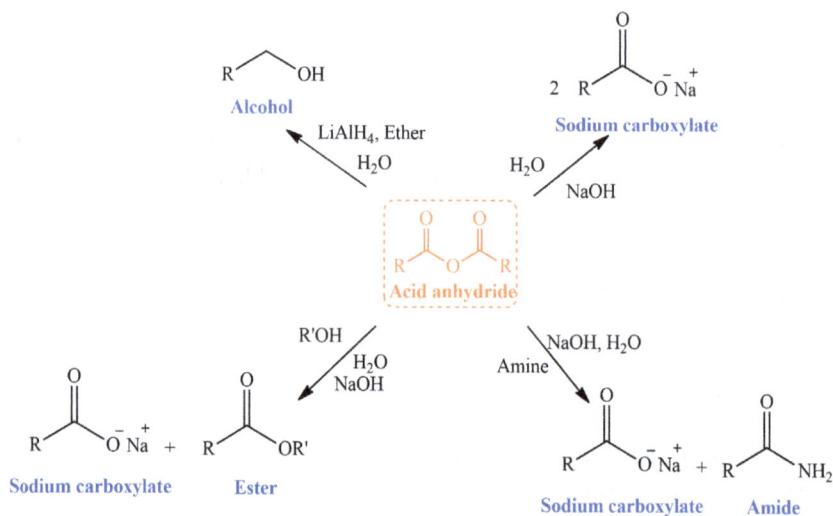

Figure 6.10: Reactions of acid anhydrides.

6.3.3.3 Reactions of Acid Chlorides

Acid chlorides react with carboxylic acids to give acid anhydrides, and with water to form carboxylic acids, with alcohols to form esters, and with ammonia, primary, and secondary amines to form amides (Figure 6.11).

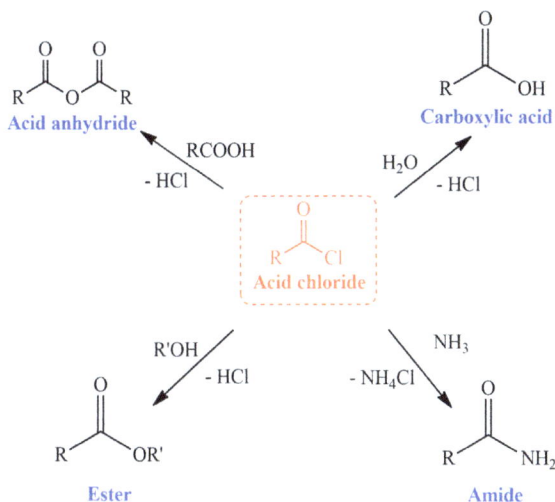

Figure 6.11: Reactions of acid chlorides.

6.3.3.4 Reactions of Amides

Amides react with water under acidic or basic conditions to give the corresponding carboxylic acids through nucleophilic acyl substitution. The primary amides react with thionyl chloride ($SOCl_2$) to produce nitriles and are also reduced by $LiAlH_4$ to give the corresponding amine. The reaction of amides a large excess of an alcohols give esters (Figure 6.12).

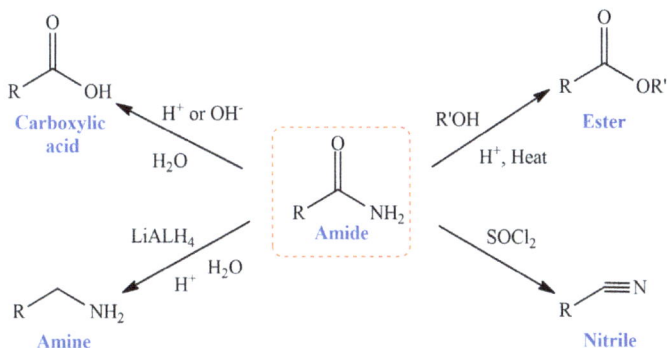

Figure 6.12: Reactions of amides.

6.3.3.5 Reactions of Nitriles

Nitriles react with LiAlH$_4$ to give imine salts that can be converted to amines by the addition of water. Nitriles react with the Grignard reagent to form imine salts. These salts can then be hydrolyzed by water to form ketones. They are also converted by acids to amides and carboxylic acids (Figure 6.13).

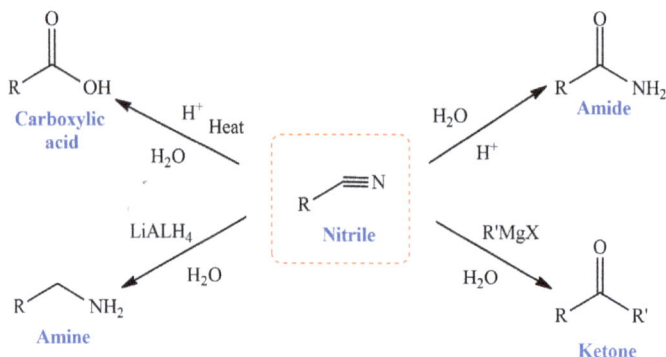

Figure 6.13: Reactions of nitriles.

6.4 Essential Terms

Acid chloride A compound that is derived from carboxylic acid by replacement of hydroxyl by chlorine.

Acyl compounds Compounds that have RCO- or ArCO- groups.

Amide An organic compound containing RCONH$_2$ group and derived from carboxylic acid by replacement of hydroxyl by amino group.

Carbonyl compounds Compounds that have CO- group.

Carboxylic acids Carbonyl compounds that have two functional groups namely the carbonyl and the hydroxyl groups.

Carboxylic acid derivatives Classes of organic compounds derived from carboxylic acids, and each has an acyl group (RCO- or ArCO-) attached to a halogen, oxygen, or nitrogen atom.

Dicarboxylic acid Acyl compound that has two carboxyl groups.

Esters Organic compounds derived from corresponding carboxylic acids and alcohols.

Fischer esterification Acid-catalyzed reaction of alcohols and carboxylic acids to form esters.

LiAlH$_4$ Lithium aluminum hydride.

Long chain carboxylic acids (fatty acids) Carboxylic acids have carbon atoms that range from C12 to C26.

Mixed anhydride A carboxylic acid anhydride that has different aryl or alkyl groups.

Saponification Formation of soap from the hydrolysis of esters by water in the presence of strong acids or bases.

Succinic anhydride An organic cyclic compound with the general formula $(CH_2CO)_2O$.

Symmetrical anhydride A carboxylic acid anhydride that has same aryl or alkyl groups.

Terephthalic acid A carboxylic acid that has two carboxyl groups attached to positions 1 and 4 of the benzene ring. It is also named as benzene-1,4-dicarboxylic acid or 1,4-benzenedicarboxylic acid.

6.5 Problems

6.5.1   Write down the common names for the following carboxylic acids:
   i.   Pentanoic acid
   ii.   Hexanoic acid
   iii. Heptanoic acid
   iv.   Octanoic acid

6.5.2   Draw the chemical structures oxalic and isophthalic acids.

6.5.3   Briefly explain how ethyl acetate is prepared.

6.5.4   What is the reactivity order of carboxylic acid derivatives?

6.5.5   List down the common names of the carboxylic acids that are required for the preparation of each of the following esters:

Ester	Carboxylic acid
Methyl acetate	
Ethyl formate	
Ethyl butyrate	
Methyl propionate	

6.5.6   Write down the products of the treatment of ethanoic anhydride with methanol.

6.5.7   Draw the chemical structure of phthalic acid and show how it differs from isophthalic acid.

6.5.8   Give one example for symmetrical anhydride and another one for mixed anhydride.

6.5.9   Explain why only primary and secondary and not tertiary amides form hydrogen bonds with each other.

6.5.10   Draw the chemical structure of cyclopropyl butanoate.

6.5.11   Give the IUPAC name of the following compound:
         $CH_3CH_2CH(CH_3)CH(NH_2)COOH$

6.5.12   Which one of the following is the most reactive carboxylic acid derivative?

      i.   Amide

     ii. Acid chloride

   iii. Nitrile

6.5.13   What is the name of the compound results from heating ethyl benzoate in a large excess of methanol in the presence of acid catalyst?

6.5.14   Which product is obtained when acetamide (CH_3CONH_2) is heated in aqueous acid?

6.5.15   Draw the chemical structure of 1,2-cyclopentanedicarboxylic acid.

6.5.16   Name the following carboxylic acid:

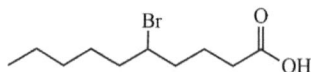

6.5.17   What is the common name of hexanedioic acid?

6.5.18   What product/s will be formed from the reaction of propanoic acid with lithium aluminum hydride?

6.5.19   Show the reaction of acetic acid with ammonia at high temperature.

6.5.20   What reducing agent is used to convert ethanamide to ethylamine?

6.5.21   Why the low-molecular weight nitriles are soluble in water?

6.5.22   What are the products of the reaction between methyl acetate and lithium aluminum hydride?

6.5.23   Explain how acetonitrile is prepared from acetamide.

6.5.24   Which product will be formed from the reaction of acetyl chloride and ammonia?

6.5.25   List down uses of the following carboxylic acids:

      i.   Malonic acid

     ii. Phthalic acid

   iii. Glutaric acid

6.5.26   Show the product of treating methyl benzoate with methylmagnesium bromide.

6.5.27   Draw the chemical structure of the following carboxylic acid derivatives:

      i.   Dimethyl oxalate

     ii. Butanoic anhydride

   iii. Octanamide

6.5.28   What is the common name of 2-amino-3-carbamoylpropanoic acid?

6.5.29   Give the IUPAC name of the linoleic acid.

6.5.30   Give one example for unsaturated long-chain carboxylic acid.

****

Chapter 7
Organic Polymers

Objectives

After studying this chapter, learners will be able to:
- Deduce notation and nomenclature of organic polymers.
- Know polymer architecture.
- Describe polymer morphology.
- Explain polymerizations reactions.
- Understand polymer stereochemistry (tacticity).
- Know copolymerization.

7.1 Notation and Nomenclature of Organic Polymers

Polymers are named by adding the prefix poly to the name of the monomer from which it is derived. The name of the monomer is kept in parenthesis when the monomer is complex or when its name is formed from two words like in poly(vinyl chloride), poly(vinyl acetate), and poly(ethylene terephthalate).

To represent polymer in an equation, for example, polypropylene (PP), the propylene monomer should be preceded by "n" and should show the C=C double bond in the structure.

The PP repeating unit is then drawn. The polymer structure is identical to the monomer except for the C=C double bond which is converted to a C–C single bond.

To show that the repeating unit is "repeated" throughout the polymer, in an end-to-end manner, a bracket is placed outside the repeating unit, and the C–C bonds are extended outside both sides of the bracket. In addition, a subscript n is placed on the bottom right corner of the polymer repeating unit as shown in Figure 7.1.

7.2 Polymers Architecture

Polymers as long-chain molecules (macromolecules) are synthesized by joining monomers through chemical reactions (formation of chemical bonds). Their architectures are quite diverse due to their high molecular weights. There are different types of polymer architecture. This includes linear and branched chains as well as comb, ladder, network, and star structures as shown in Figure 7.2.

https://doi.org/10.1515/9783111382753-007

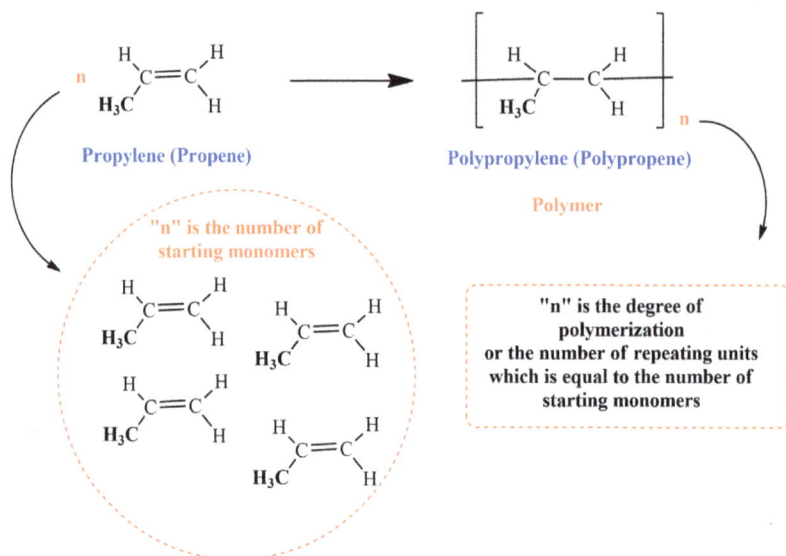

Figure 7.1: Polymer representation in an equation.

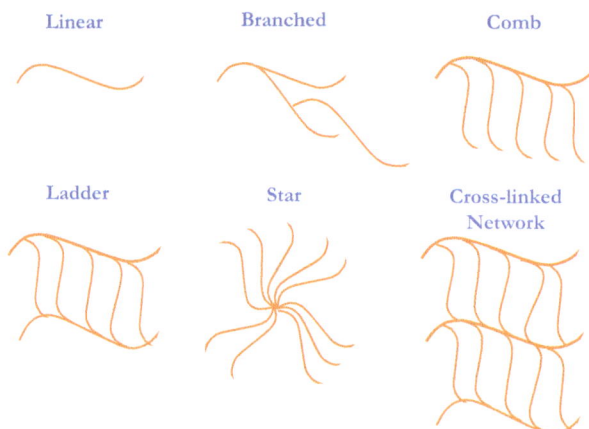

Figure 7.2: Types of polymer architecture.

7.3 Polymers Morphology

Polymer morphology refers to the overall shape of a polymer structure. Polymers, like small organic molecules, crystallize when precipitated or cooled from a melt. Inhibiting this tendency for very big molecules slows diffusion and hinders efficient chain packing.

This results in the formation of both ordered crystalline domains (crystallites) and disordered amorphous domains (Figure 7.3). The relative amounts of crystalline and amorphous domains differ from polymer to polymer and often depend on the manner in which the material is processed.

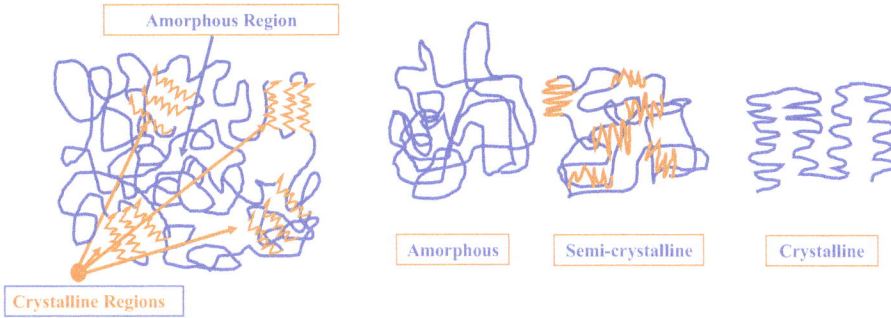

Figure 7.3: Polymer morphology.

7.4 Polymerization Reactions

A chemical reaction that converts small molecules (monomers) to large molecules (polymers) is called polymerization. There are two types of these reactions "addition and condensation." These types will be discussed in the following paragraphs and summarized in Figure 7.4.

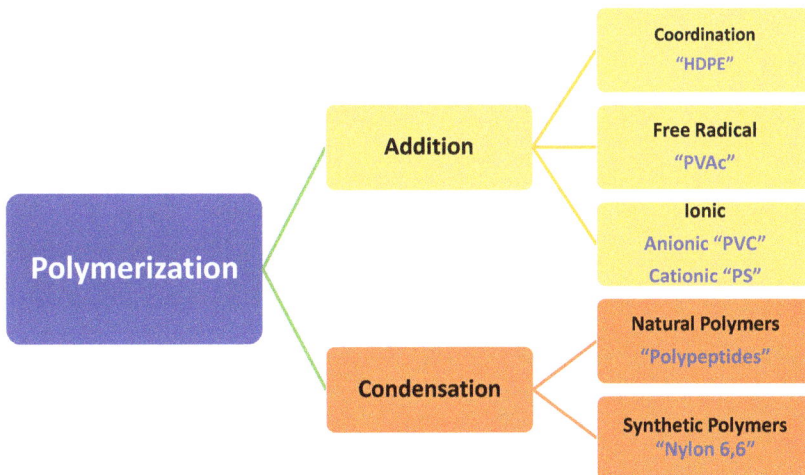

Figure 7.4: Types of polymerization reactions.

A polymerization is called an addition polymerization if the entire monomer molecule becomes part of the polymer. A polymerization is called condensation polymerization if part of the monomer molecule is released out when the monomer becomes part of the polymer. This part is usually a small molecule like water or HCl.

For example, nylon 6,6 is made from adipoyl chloride (hexanedioyl dichloride) or adipic acid (hexanedioic acid) and hexamethylene diamine. The chlorine atoms from the adipoyl chloride along with one of the amine hydrogen atoms are released as HCl gas or the hydroxyl group from the adipic acid along with one of the amine hydrogen atoms are released as H_2O as shown in Figure 7.5.

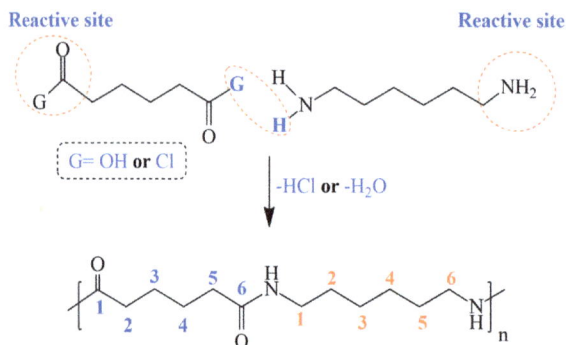

Figure 7.5: Condensation polymerization between hexanedioyl dichloride (or adipic acid) and hexamethylene diamine.

In a chain growth polymerization, monomers become part of the polymer one at a time. All the monomers from which addition polymers are made are alkenes or functionally substituted alkenes as shown in Figure 7.6.

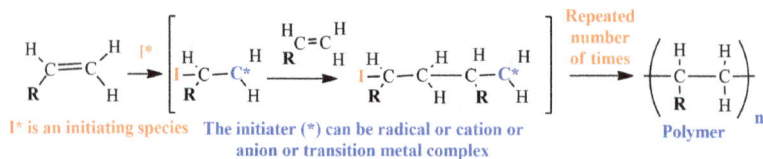

Figure 7.6: Addition polymerization.

The types of chain-growth polymerization are:
- Radical polymerization: The initiator is a radical, and the propagating site of reactivity is a carbon radical.
- Cationic polymerization: The initiator is an acid, and the propagating site of reactivity is a carbocation.

– Anionic polymerization: The initiator is a nucleophile, and the propagating site of
reactivity is a carbanion.
– Coordination catalytic polymerization: The initiator is a transition metal complex,
and the propagating site of reactivity is a terminal catalytic complex.

7.5 Polymer Stereochemistry (Tacticity)

The relative configurations of polymers' chiral centers are important in determining
the properties of a polymer. Polymers with identical configurations at all chiral centers along the chain are called isotactic polymers.

Those with alternating configurations are called syndiotactic polymers, and those
with completely random configurations are called atactic polymers (Figure 7.7).

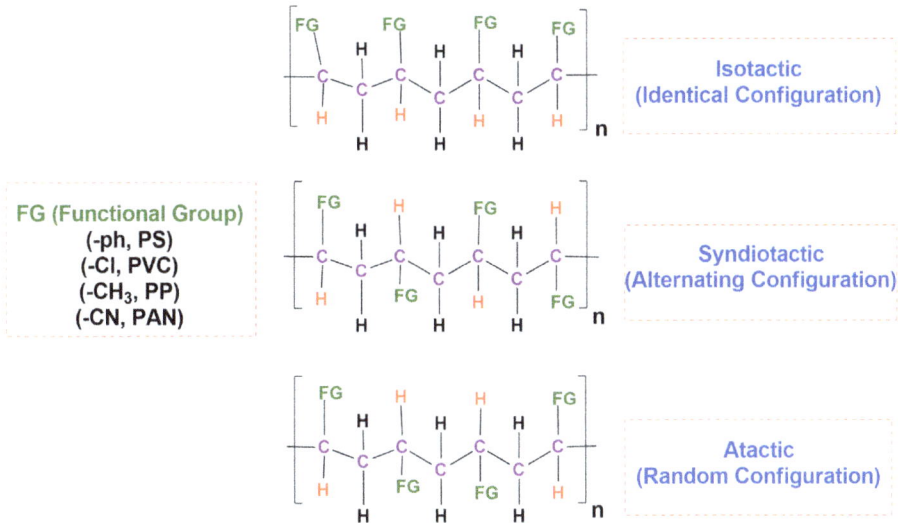

Figure 7.7: Polymer stereochemistry.

7.6 Commodity Polymers

Commodity polymers or commodity plastics are plastics that are produced in high
quantities for different daily uses. Selected examples are listed in Table 7.1.

Table 7.1: Selected examples of commodity polymers.

Polymer (abbreviation)	Repeating unit (monomer)
Polyethylene (PE)	H H $\|$ $\|$ —C—C— $\|$ $\|$ H H
Poly (vinyl chloride) (PVC)	H H $\|$ $\|$ —C—C— $\|$ $\|$ H Cl
Polypropylene (PP)	H H $\|$ $\|$ —C—C— $\|$ $\|$ H CH$_3$
Polystyrene (PS)	H H $\|$ $\|$ —C—C— $\|$ $\|$ H C$_6$H$_5$
Poly (methyl methacrylate) (PMMA)	H CH$_3$ $\|$ $\|$ —C—C— $\|$ $\|$ H COOCH$_3$

7.7 Copolymers

When two or more monomers are polymerized together, the formed molecule is called copolymer. They are classified into

- Random – **A** and **B** are positioned along chain randomly.

  (A) (A) (B) (A) (B) (B) (A) (B) (A)

- Alternating – **A** and **B** are positioned in an alternate order within the chain.

  (A) (B) (A) (B) (A) (B) (A) (B) (A)

- Block – large blocks of **A** units alternate with large blocks of **B** units.

  (A) (A) (A) (B) (B) (B) (A) (A) (A)

– Graft – chains of **B** units grafted onto **A** backbone.

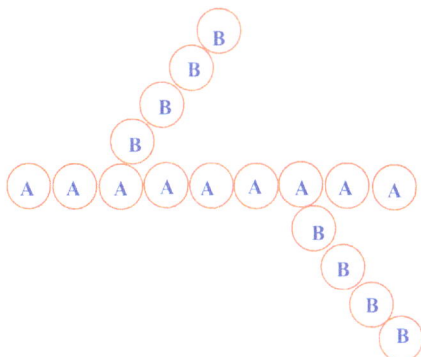

7.8 Essential Terms

Addition polymers Polymers that result from the addition of monomers to the growing chain.

Alternating copolymer A copolymer with repeating units positioned in an alternate order within the chain.

Anionic polymerization A type of polymerization in which anionic initiator transfers an electron (or negatively charged group) to a monomer to become reactive.

Atactic polymer A polymer has substituents on random sides of the backbone.

Block copolymer A copolymer with large blocks of different repeating units alternate within the chain.

Branch chain polymers Polymers that have short chains that branching at random points from the main linear chain.

Cationic polymerization A chain growth polymerization in which monomers are activated through a transfer of charge by a cationic initiator.

Chain reaction Reaction that involves macroradicals and proceeds by stepwise mechanism.

Commodity polymers or commodity plastics Plastics that are produced in high quantities for different daily uses.

Condensation polymers Polymers that are formed by the combination of two bifunctional monomers and releasing of small molecules.

Copolymer A polymer made from more than one type of monomers that have similar or different physical properties.

Cross-linked or network polymers Polymers that have monomers connected together to form a 3-D structure or network.

Degree of polymerization A number of repeating units connected together by chemical bonds.

Graft copolymer A copolymer with one type of chains grafted onto a backbone of another type.

Isotactic polymer A polymer has substituents on the same side of the backbone.

Linear polymers Polymers that form long straight chains of identical links connected together.

Polymers Polymers are macromolecules made of a number of repeating chemical units joined together by chemical bonds.

Polymerization A process by which small molecules (monomers) are converted to large molecules (polymers).

Random copolymer A copolymer with repeating units positioned along chain randomly.

Syndiotactic polymer A polymer has substituents on alternating sides of the backbone.

Vulcanization A chemical process in which elastomers cross-link by heating with sulfur.

7.9 Problems

7.9.1 What kind of monomers is required for the condensation polymerization to occur?

7.9.2 What kind of initiator is involved in the radical polymerization?

7.9.3 How are substituents arranged in the atactic polymers?

7.9.4 How are polymers classified on molecular forces basis?

7.9.5 How are substituents positioned in the isotactic polymers?

7.9.6 What is the difference in chain structure between polyethylene and propylene?

7.9.7 What are the two reactants required to make the Novolacs?

7.9.8 What are the main differences between addition polymers and condensation polymers?

7.9.9 Draw the chemical structure of the repeating unit that required to make polystyrene.

7.9.10 Draw the structure of poly(acrylonitrile).

7.9.11 What is the chemical structure of Teflon?

7.9.12 Give an example of one polymer which readily formed by cationic polymerization.

7.9.13 What kind of polymer tacticity in which the substituents are positioned randomly throughout the polymer backbone?

7.9.14 Draw the chemical structure of atactic polyethylene.

7.9.15 Draw the chemical structure of isotactic polystyrene.

7.9.16 Which element is needed for the vulcanization of rubber?

7.9.17 Is PVA a condensation or addition polymer?

7.9.18 What is the name of substance that is added to a polymer to lower its T_g?

7.9.19 What kind of stereochemistry (tacticity) do the following polymers have?

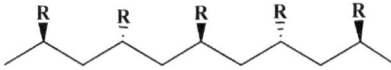

7.9.20 What is the full name of the PMMA polymer?

7.9.21 What are the polymer stabilizers?

7.9.22 List down three types of polymer architecture.

7.9.23 What is the molecular weight of the polyvinyl acetate polymer whose degree of polymerization (DP = n) is equal 500?

7.9.24 What are the differences between random and alternating copolymers?

7.9.25 Give two examples for the catalysts that are used in cationic polymerization.

7.9.26 Which one of following polymers is a copolymer?
   i.   Styrene-butadiene
   ii. Polyvinyl acetate
   iii. Polypropylene

7.9.27 The polyvinyl chloride (PVC) polymer is made from the following monomer:
   i. CH_3CH_2Cl
   ii. $CH_2=CCl_2$
   iii. $CH_2=CHCl$

7.9.28 A polymerization reaction that produces a polymer and releases at the same time small molecules is known as:
   i.   Addition
   ii. Condensation
   iii. Elimination

7.9.29 Polystyrene polymers formed from the polymerization of
   i. $CH_3CH=CH_2$
   ii. $C_6H_5-CH=CH_2$
   iii. $CH_2=CH-CH_2CH_3$

7.9.30 Name and draw the monomer from which the Teflon polymer is made.

****

Chapter 8
Stereochemistry Topics and Concepts

Objectives

After studying this chapter, learners will be able to:
– Know the concept of stereochemistry.
– Understand isomerism.
– Describe chirality and achirality.
– Explain characteristics of homologous series.
– Distinguish between the *R-S* and *E-Z* notational systems.
– Know meso compounds.
– Distinguish between syn-anti isomers and cis-trans isomers.
– Understand tautomerism and epimerization

8.1 Stereochemistry and Isomers

Stereochemistry is the study of the three-dimensional (3-D) aspects of the molecules. Isomers stand for a Greek word in which **iso** means equal and **mers** mean parts. They are compounds that have the same formula but different structures. They can also be defined as molecules that have the same number and types of atoms, the same formula, but are different. There are various types of isomers. A simplified overview is summarized in Figure 8.1.

Structural isomers (constitutional isomers) have their atoms joined together in a different order (different chemical bonding). They are very common in organic chemistry. Examples of structural isomers are skeletal (have different carbon skeleton), functional (have different functional groups), and positional (their functional groups are in different positions).

Stereoisomers are joined together in the same order but differ only in the arrangement of their atoms in space. Examples of stereoisomers are optical isomers (enantiomers) and geometrical isomers. Optical isomers are non-superimposable (cannot lay on each other), mirror-image isomers with identical physical properties except for the direction that they rotate the plane of the polarized light. Geometrical isomers (diastereoisomers) are non-superimposable, non-mirror-image isomers.

They differ in the geometry of the groups on the double bond, where the substituents are on the same side in *cis* and are on the opposite side of double bond in *trans*.

https://doi.org/10.1515/9783111382753-008

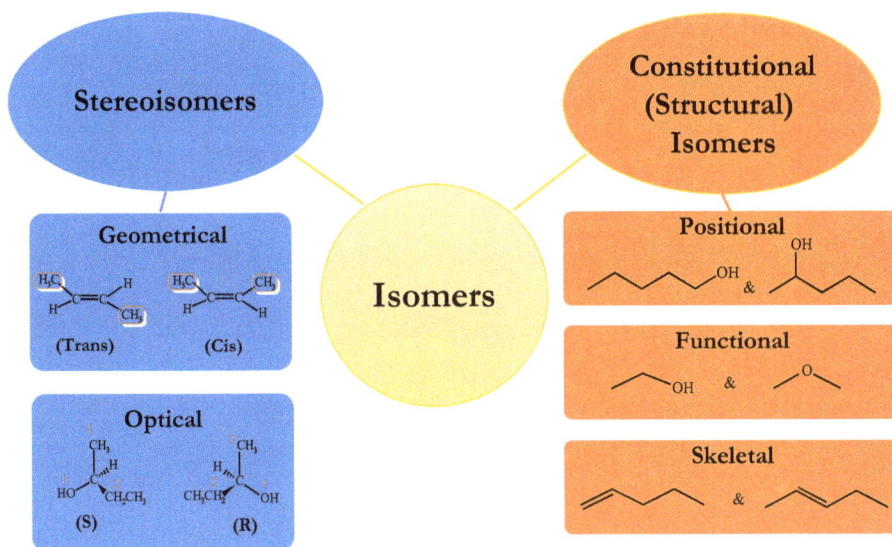

Figure 8.1: Different types of isomers.

8.2 The Homologous Series

8.2.1 Definition

A series of related compounds that have a similar structure, but each member differs from the adjacent member of the series by a methylene group (Figure 8.2).

CH_3-OH, CH_3-CH_2-OH, $CH_3CH_2CH_2$-OH,
These are members of a homologous series called " Alcohols"

CH_3-SH, CH_3-CH_2-SH, $CH_3CH_2CH_2$-SH,
These are members of a homologous series called " Thiols"

CH_3-COOH, CH_3-CH_2-COOH, CH_3-CH_2-CH_2-COOH,
Thes are members of a homologous series called "Carboxylic acids"

CH_3-O-CH_3, CH_3CH_2-O-CH_3, CH_3-CH_2-CH_2-O-CH_3,
These are members of a homologous series called " Ethers"

CH_3-CHO, CH_3-CH_2-CHO, CH_3-CH_2-CH_2-CHO,
These are members of a homologous series called "Aldehydes"

Figure 8.2: Different homologous series.

8.2.2 Characteristics of Homologous Series

- The molecular formulas of different members of the series differ from the next members by a $-CH_2-$ group.
- All members of the series contain the same functional group.
- , All members of the series have the same chemical reactivity because they have the same functional group.
- Physical properties of the series vary with changes in the number of carbon atoms, that is, with an increase in molecular weight.
- General methods of preparation apply to any member of the series.

8.3 Molecular Chirality

Molecular chirality is important because of its applicability to stereochemistry in organic chemistry, inorganic chemistry, physical chemistry, biochemistry, and supramolecular chemistry.

The asymmetry or handedness of things or molecules and the inability of certain objects or molecules to be overlaid on their mirror images are both referred to as chirality (Figure 8.3). The expression is derived from the Greek word "kheir," which means "hand."

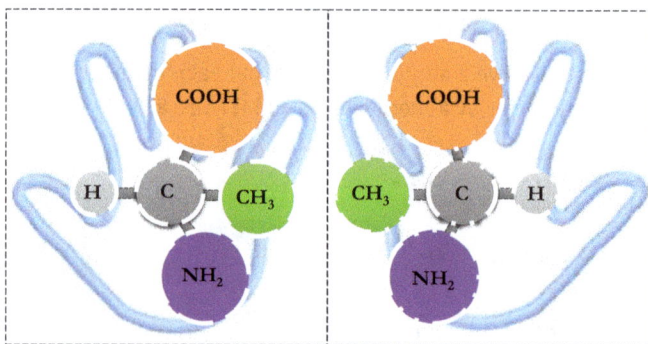

Figure 8.3: Chirality or handedness.

If a molecule's two mirror-image forms cannot be superimposed in three dimensions, it is **chiral** (Figure 8.4). If the tetrahedral carbon has four distinct substituents (atoms), the molecule is also **chiral**.

An **achiral** molecule has two mirror-image forms that are superimposable in three dimensions (Figure 8.5).

Also, the absence or presence of a plane of symmetry or internal symmetry is often a deciding feature of the existing or lacking chirality in molecules.

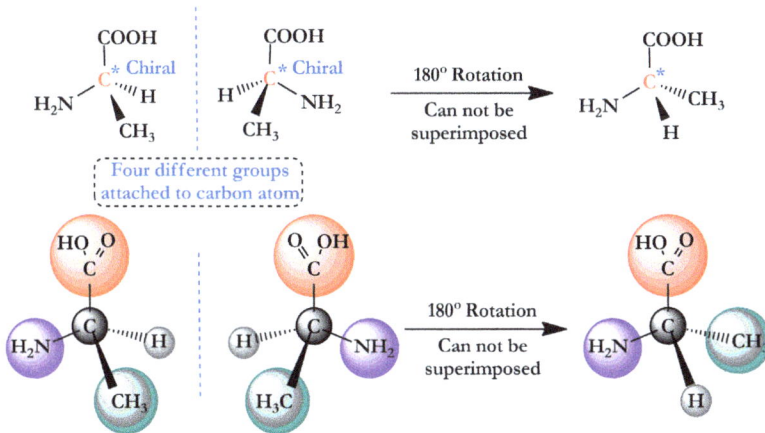

Figure 8.4: Chiral compounds.

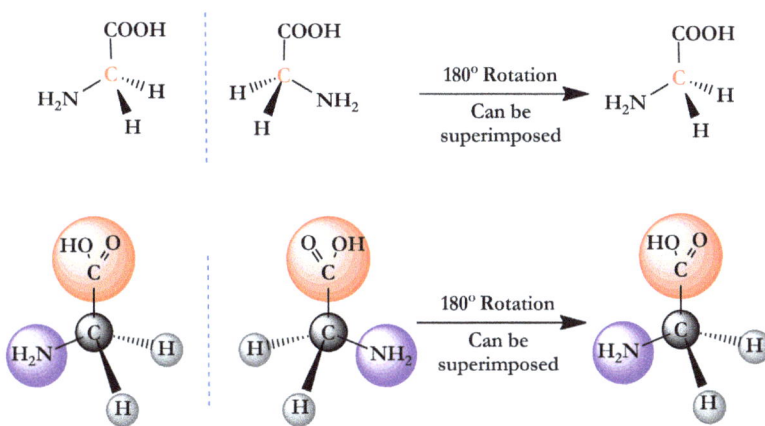

Figure 8.5: Achiral compounds.

Achiral objects possess a plane of symmetry or internal symmetry, allowing them to be superimposed onto their mirror images. In other words, achiral objects lack handedness and are identical to their mirror images.

On the contrary, chiral objects lack this internal symmetry and cannot be superimposed onto their mirror images. This characteristic is a fundamental property of chirality.

Our hands are chiral, which means that they lack a plane of symmetry or internal symmetry. It is impossible to line up the left and right hands exactly. The hand cannot be divided so that one side mirrors the other. For example, propanoic acid has a plane of symmetry when lined up, making one side of the molecule a mirror image of the other, and is achiral, but lactic acid has no plane of symmetry and is chiral, as seen in Figure 8.6.

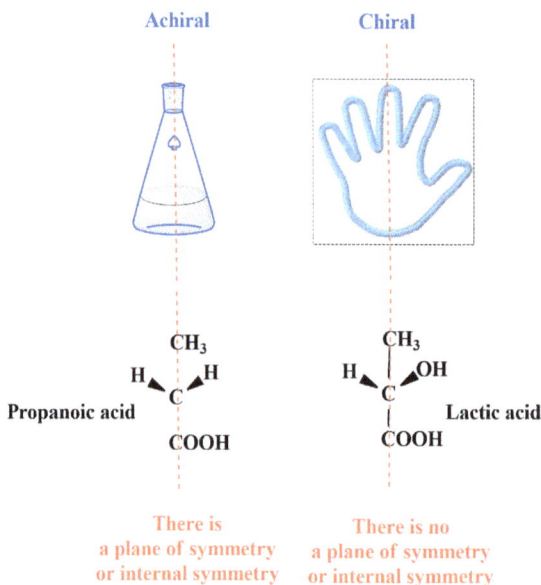

Figure 8.6: Absence or presence of a plane of symmetry or internal symmetry.

8.4 Stereocenter and Chiral Center

The primary distinction between a stereocenter and a chiral center is that a stereocenter is any point of a molecule that can give a stereoisomer when two groups at this point are interchanged, whereas a chiral center is an atom in a molecule that can give an enantiomer when two groups at this center are interchanged. All chiral centers are stereocenters; however, all stereocenters are not chiral centers.

The key differences between stereocenter and chiral center are summarized in Table 8.1.

8.5 The E-Z Notational System

If atoms of a higher atomic number are on the same side of the double bond, the configuration is **Z** (**Z** stands for the German word zusammen means together), but if atoms with a higher atomic number are on the opposite sides of the double bond the configuration is **E** (**E** stands for German word entgegen means opposite; Figure 8.7).

Table 8.1: Stereocenter versus chiral center.

Stereocenter	Chiral center
– A stereocenter is a point in a molecule or sp^2-/ sp^3-hybridized carbon atom. – Stereocenter has either three or four groups attached to it. – Stereocenter has either single bonds or double bonds around it. – The interchange of groups at the stereocenter forms stereoisomers.	– Chiral center is a sp^3-hybridized carbon atom to which four different atoms or groups of atoms are bonded. – Chiral center has only single bonds around it. – The interchange of groups at the chiral center forms enantiomers.

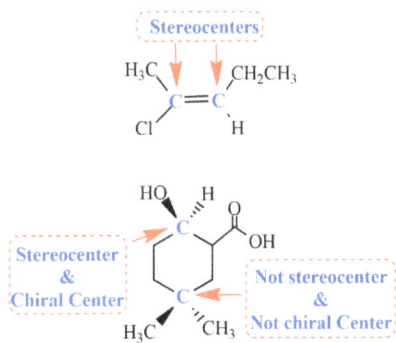

8.6 The R-S Notational System

According to Cahn-Ingold-Prelog sequence rules, the absolute configuration (R & S) can be assigned as follows:
- Substituents (or atoms) at the stereogenic center are ranked according to their atomic numbers. The lowest-ranked substituent is pointed away.
- If the order of decreasing precedence of the highest-ranked substituents appears in a clockwise sense, the absolute configuration is R (**R stands for the Latin word rectus means right**); if the order is counterclockwise, the absolute configuration is S (**S stands for the Latin word sinister means left**; Figure 8.8).

8.7 Racemic Mixture

A racemic mixture, often known as a "racemate," is defined by IUPAC as a 50:50 or 1:1 equimolar mixture of two enantiomers. There won't be an optically inactive net rotation since each enantiomer rotates plane-polarized light to an equal and opposite degree because they are present in equal amounts.

Opposite sides   Same side

$$H^1 \quad C^6$$
$$O^8$$
$$Br^{35}$$

Lower   Higher   Higher   Higher

H   Br   HO   Br

C=C   C=C

HO   CH$_3$ H   CH$_3$

Higher   Lower   Lower   Lower

E-Configuration   **Z-Configuration**

Opposite sides   Same side

$$H^1 \quad C^6$$
$$O^8$$
$$Cl^{17}$$
$$Br^{35}$$

Lower   Higher   Higher   Higher

Cl   CHO   Br   CHO

C=C   C=C   OOH

Br   CH$_2$OH   Cl   CH$_2$OH OHHH

Higher   Lower   Lower   Lower

E-Configuration   **Z-Configuration**

Opposite sides   Same side

$$H^1 \quad F^9$$
$$Cl^{17}$$
$$Br^{35}$$

Higher   Higher   Higher

Cl   Lower   Cl   Br

                F

C=C   C=C

H   Br   H   F

Lower   Higher   Lower   Lower

E-Configuration   **Z-Configuration**

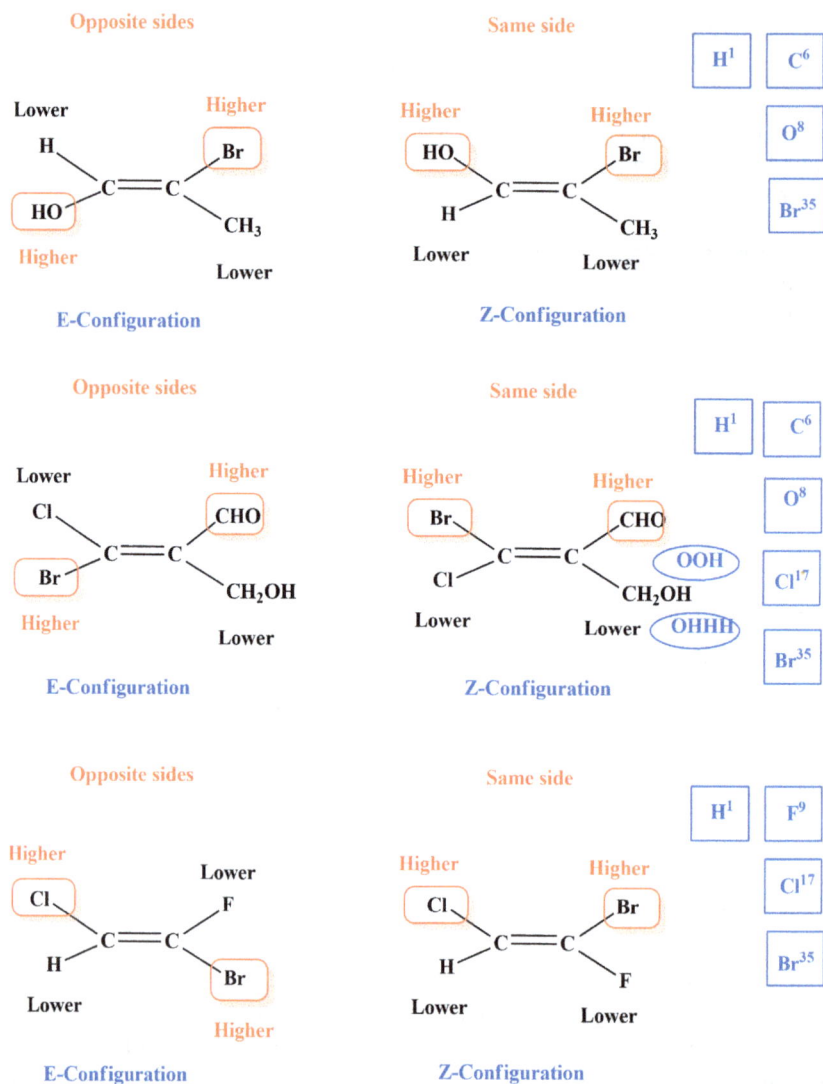

Figure 8.7: The *E-Z* notational system.

A mixture cannot be characterized as "racemic" until all the above-mentioned three parameters are met. A racemic mixture marked by (+/–) or (±) to signify equal proportions of the (+) (dextrorotatory, rotates plane-polarized light clockwise) and (–) (levorotatory, rotates plane-polarized light counterclockwise) enantiomers. As illustrated in Figure 8.9, a (+/–) can be assigned to the two enantiomers in the racemic mixture of (+/–) butan-2-ol, the racemic mixture of (+/–) lactic acid, and the racemic mixture of (+/–) tartaric acid.

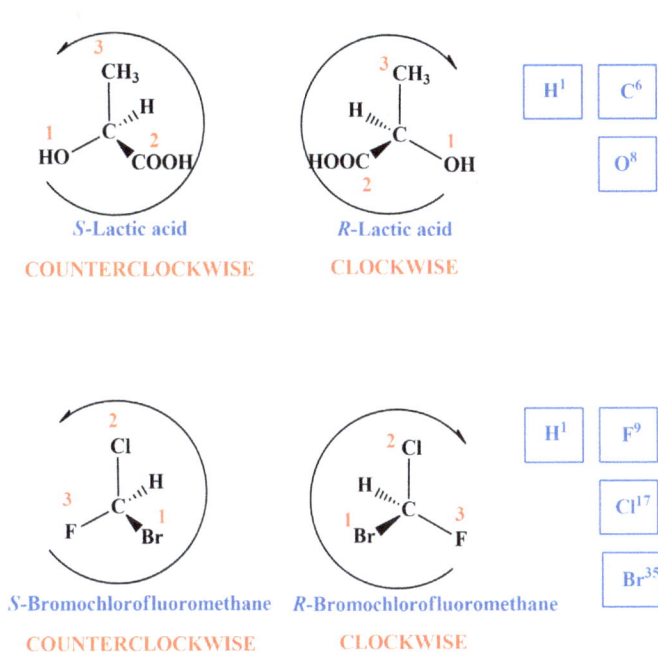

Figure 8.8: The *R-S* notational system.

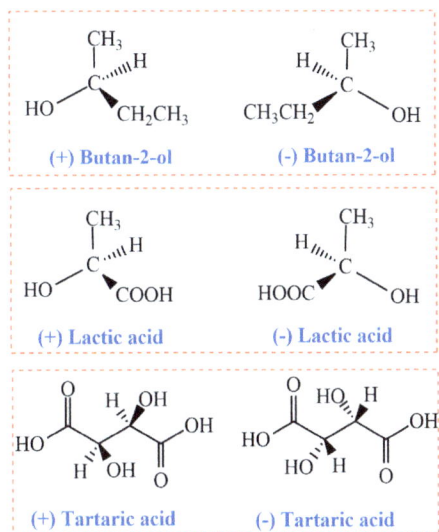

Figure 8.9: Racemic mixture of (+/−) butan-2-ol, (+/−) lactic acid, and (+/−) tartaric acid.

A racemic mixture is created by combining an equal number of moles of two enantiomers. Figure 8.10 shows how to make 2.0 moles of a racemic mixture of lactic acid from 1.0 mol of (–) lactic acid and 1.0 mol of (+) lactic acid.

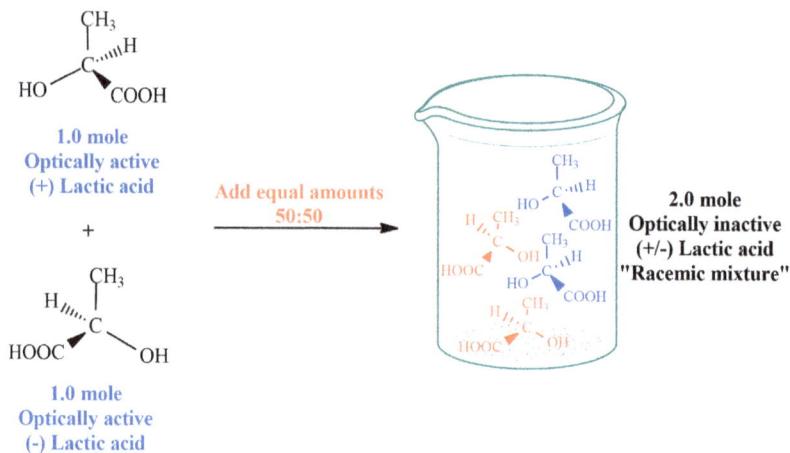

Figure 8.10: Racemic mixture of equimolar mixture (50:50) of two enantiomers of lactic acid.

8.8 Meso Compounds

Meso compounds are achiral compounds with a number of chiral centers. They are optically inactive and possess an internal plane of symmetry that allows it to be superimposable on its mirror image. They are distinguished by having two comparable halves, as seen in Figure 8.11.

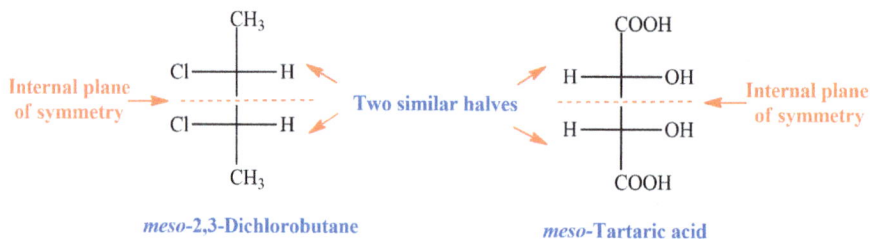

Figure 8.11: Examples of meso-compounds.

8.9 D-L Isomers

To denote the configuration of amino acids and sugars, the stereoscriptors **D-** (from Latin dexter, right) and **L-** (from Latin Laevus, left) are used. The descriptors **D-** and **L-** are written in small letters and separated from the remainder of the name by a hyphen.

The molecule's configurational stereochemistry is denoted by its **D-** and **L**-enantiomers. In carbohydrates or sugars, the hydroxy group is attached to the left side of the asymmetric carbon that is farthest from the carbonyl group in **L**-isomers, whereas the hydroxy group is attached to the right side in **D**-isomers. However, in amino acids the hydroxy group is attached to the left side of the chiral carbon in **L**-isomers and to the right side in the **D**-isomers as illustrated in Figure 8.12.

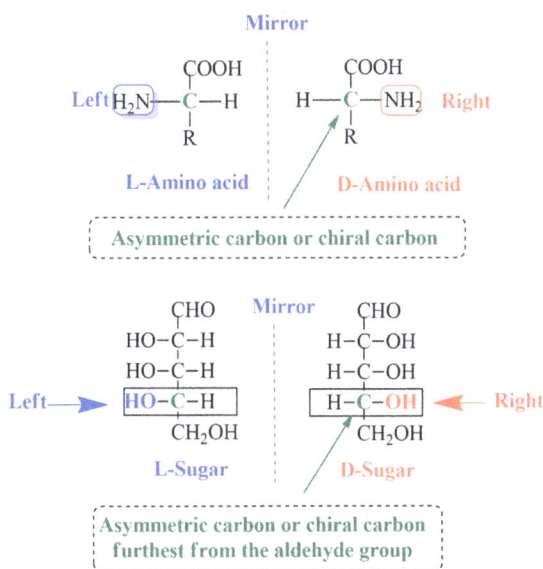

Figure 8.12: Examples of D- and L-isomers.

8.10 Epimers and Epimerization

Epimers are sugars that differ in a single location in the arrangement of the –OH groups. An epimer is one of two diastereomers in stereochemistry. Only one stereogenic center out of at least two has an opposing configuration for the two epimers. The molecules' other stereogenic centers are identical. The interconversion of one epimer to the other epimer is known as epimerization.

For example, **D**-idose and **D**-talose epimers create a single difference at **C-3** carbon. On the other hand, **D**-allose and **D**-altrose epimers create a single difference at **C-2** carbon, as shown in Figure 8.13.

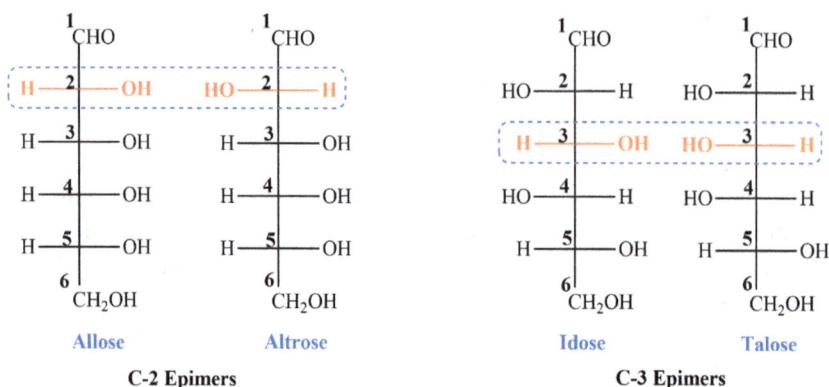

Figure 8.13: Examples of epimers.

8.11 Syn-Anti Isomers

Geometrical isomerism is denoted by the prefixes syn and anti. Both H and OH are present on the same side of the double bond in syn, while H and OH are present on opposite sides of the double bond in anti. The same is true in nucleic acid chemistry, where the nucleic acid base can exist in two different orientations around the N-glycosidic linkage. As seen in Figure 8.14, these conformations are characterized

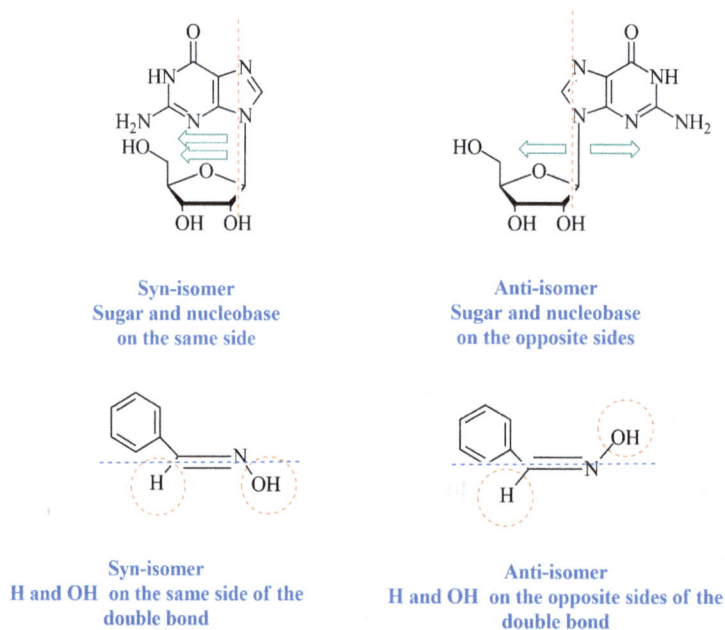

Figure 8.14: Examples of syn- and anti-isomers.

as syn, which indicates that the sugar and nucleobase are on the same side, and anti, which shows that the sugar and nucleobase are on different sides.

8.12 Cis-Trans Isomers

Cis isomers are compounds that have two comparable atoms on the same side of a molecule's double bond. Trans isomers are compounds that have two comparable atoms on opposing sides of a double bond. Figure 8.15 shows some examples.

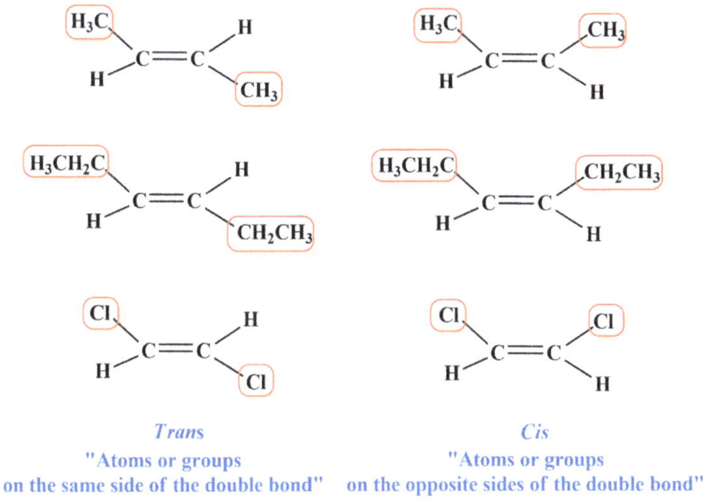

Trans
"Atoms or groups
on the same side of the double bond"

Cis
"Atoms or groups
on the opposite sides of the double bond"

Figure 8.15: Examples of cis- and trans-isomers.

8.13 Exo-Endo Isomerism

The prefixes exo (outside or away or outwards) and endo (inside or into or toward) define the relative configuration of bridging bicyclic molecules. Exo or endo is determined by the position of a substituent in the main ring relative to the shortest bridge. When the substituent is facing the bridge, the exo description is assigned to it. It is endo-configured while facing away from the bridge, as shown in Figure 8.16.

8.14 Tautomerism

The presence of two or more chemical compounds that may easily interconvert, often by merely transferring a hydrogen atom between two other atoms with which it estab-

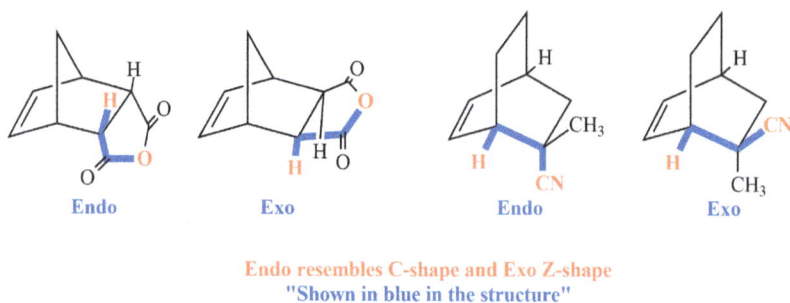

Endo resembles C-shape and Exo Z-shape
"Shown in blue in the structure"

Figure 8.16: Examples of exo- and endo-isomers.

lishes a covalent connection, is referred to as tautomerism. Tautomeric compounds, unlike other classes of isomers, exist in dynamic equilibrium with each other.

The most common type of tautomerism is that involving carbonyl, or keto, compounds and unsaturated hydroxyl compounds, or enols; the structural change is the shift of a hydrogen atom between atoms of carbon and oxygen, with the rearrangement of bonds. This is known as keto-enol tautomerism.

In many aliphatic aldehydes and ketones, such as acetaldehyde, the keto form is prominent, while in phenols, the enol form, which is stabilized by the aromatic feature of the benzene ring, is the prominent form. Purine and pyrimidine nucleotide bases can also exist as aromatic heterocyclic hydroxyl pyrimidine or purine tautomers. Despite the increased aromatic ring stability, these bases prefer amide-like structures, as illustrated in Figure 8.17.

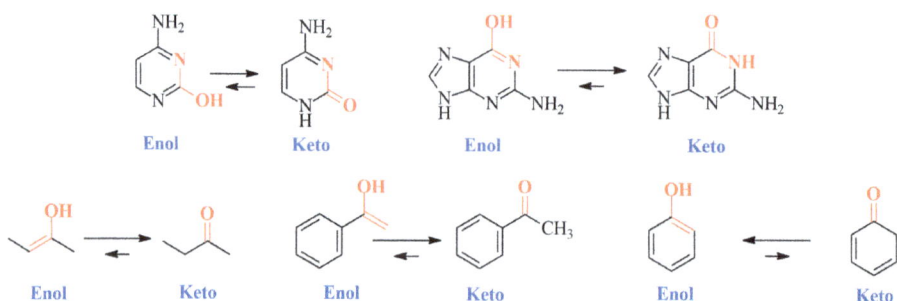

Figure 8.17: Examples of keto-enol tautomerism.

8.15 Essential Terms

Asymmetric carbon atom (chiral carbon atom) A carbon atom that is bonded to four different groups.

Cahn-Ingold-Prelog The method for designating the absolute configuration of a chiral center as either (R) or (S).

Chiral center An atom in a molecule that can give an enantiomer when two groups at this center are interchanged.

Chirality The asymmetry or handedness of things or molecules and the inability of certain objects or molecules to be overlaid on their mirror images.

Cis-trans isomers Diastereomers that differ in their cis-trans arrangement on a ring or double bond.

Diastereomers Stereoisomers that are not mirror images of each other.

Enantiomers Mirror-image isomers.

Epimerization Interconversion of one epimer to the other epimer.

Homologous series A series of related compounds that have a similar structure, but each member differs from the adjacent member of the series by a methylene group.

Isomers Different compounds with the same molecular formula.

Meso compounds Achiral compounds that have asymmetric carbon atoms, but their molecules are characterized by having two similar halves.

Racemic mixture The mixture that have equal amounts of two enantiomers.

Stereocenter Any point of a molecule that can give a stereoisomer when two groups at this point are interchanged.

Stereochemistry The study of the three-dimensional structure of molecules.

Stereoisomers Isomers whose atoms are bonded together in the same order but differ in the special arrangement.

Structural isomers also known as constitutional isomers Isomers that differ in the order in which their atoms are bonded together.

Tautomerism The presence of two or more chemical compounds that may easily interconvert.

8.16 Problems

8.16.1   Explain the difference between cis-trans diastereoisomers and give one example.

8.16.2   Give one example for homologues series.

8.16.3   Draw the skeletal structures of 2-methylpropane and butane. Are these compounds isomers and if yes what type of isomerism?

8.16.4   Which one of the following structures has **Z** configuration and why? (Atomic numbers for Cl = 17, F = 9, H = 1, Br = 35)

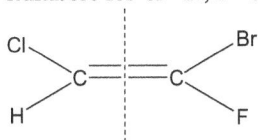

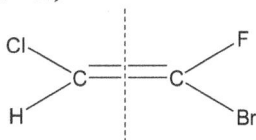

8.16.5 Show the chiral center in the following chemical structure:

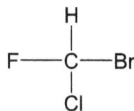

$$\text{F} - \overset{\displaystyle H}{\underset{\displaystyle Cl}{C}} - \text{Br}$$

8.16.6 Circle the compound having **E** configuration according to **E-Z** notational system. (Atomic number: I = 53; Cl = 17; F = 9 and H = 1)

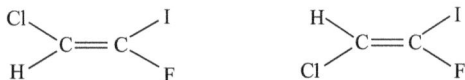

8.16.7 Name the three types of the structural isomers.

8.16.8 Circle the compound having **S** configuration according to **R-S** notational system. (Atomic Number: Br = 35; Cl = 17; F = 9; and H = 1)

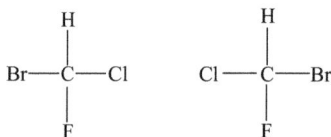

8.16.9 Show the stereocenters in the following chemical structures:

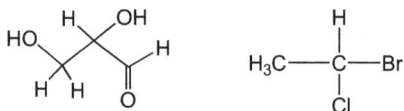

8.16.10 Can aldehydes be considered as homologous series? Explain.

8.16.11 A series of compounds, like the aldehydes, that differ only by the number of $-CH_2-$ groups, is called a_____ series.

8.16.12 Are the following compounds cis-trans isomers or constitutional isomers?

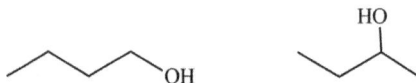

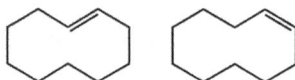

8.16.13 Are the following compounds cis-trans isomers or constitutional isomers?

8.16.14 Give an example for functional constitutional isomers.

8.16.15 Assign the following structures as cis- or -trans diastereoisomers.

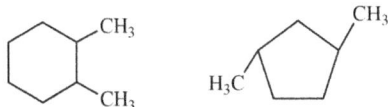

8.16.16 Which one of the following molecules is chiral?

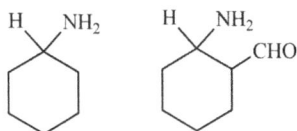

8.16.17 Glycine is an amino acid found in proteins is achiral. Why?

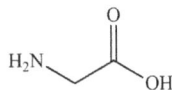

8.16.18 Assign **R** or **S** configuration to each of the following molecules:

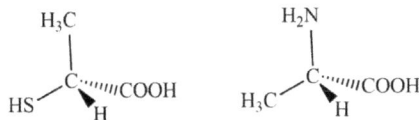

8.16.19 Which of the following structures represent meso compounds?

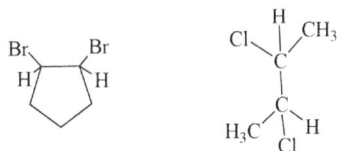

8.16.20 Do the members of the homologous series have the same functional groups?

8.16.21 Determine if the following molecules are meso.

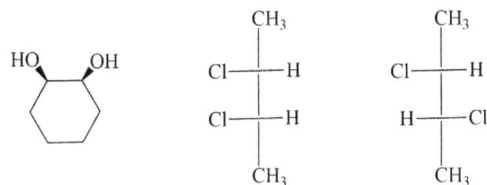

8.16.22 What is a racemic mixture?

8.16.23 Does the following structure represent a meso compound?

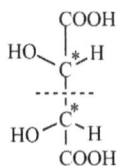

8.16.24 What is the prominent form or tautomer in phenols?

8.16.25 Give one example for the keto-enol tautomerism.

8.16.26 What is the ration of the equimolar mixture of two enantiomers in the racemic mixture?

8.16.27 Give an example for C-4 epimers.

8.16.28 Define achiral molecules.

8.16.29 Would it be possible to make racemic mixture?

8.16.30 What do the prefixes Exo or endo stand for?

****

Chapter 9
The Most Common Organic Reactions

Objectives

After studying this chapter, learners will be able to:
- Understand reactions involving the carbonyl group.
- Know reactions involving C–C double and triple bonds.
- Recognize reactions involving alkyl and aryl halides.
- Describe reactions involving alcohols.

There are numerous organic reactions that occur under various reaction conditions and involve various functional groups. These conditions range from severely cold to harsh and high-temperature extremes. Among these are the ones that involve the carbonyl group, carbon-carbon double and triple bonds, alkyl and aryl halide groups, and alcohol functional groups. The following tables summarize a handful of the most typical reactions using certain functional groups.

9.1 Carbonyl Group Involved Organic Reactions

Reactions that involve carbonyl groups are very important and play a crucial role in making desirable organic products that have and still have their role in many applications. Among those reactions are the Nozaki-Hiyama-Kishi reaction, the Forster reaction, the Baylis-Hillman reaction, the Dieckmann condensation, the Clemmensen reduction, the Wolff-Kishner reduction, the Emmert reaction, the Claisen condensation, the Vilsmeier-Haack reaction, and the Thiele reaction. All these reactions are briefly discussed and presented in Table 9.1.

9.2 Organic Reactions Involve Carbon-Carbon Double and Triple Bonds

Carbon-carbon double and triple-bond reactions are extremely important and play an important role in the creation of desired organic compounds that have and continue to play a role in a variety of applications. The Simmons-Smith reaction, ozonolysis, alkyne zipper reaction, Bergman cyclization, Brown hydroboration, Cope rearrangement, Glaser-Eglinton-Hay coupling, Kucherov reaction, oxymercuration-reduction, and Upjohn dihydroxylation are among these reactions. Table 9.2 briefly discusses all of these reactions.

https://doi.org/10.1515/9783111382753-009

Table 9.1: Selected reactions that involve carbonyl group.

1. Nozaki-Hiyama-Kishi Reaction

The Nozaki-Hiyama-Kishi reaction is a nickel-chromium coupling that forms an alcohol from the reaction of an aldehyde with an allyl or vinyl halide.

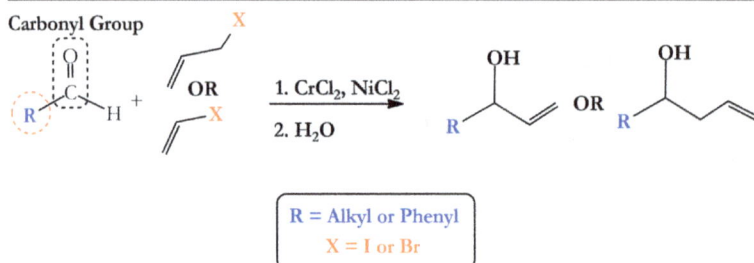

2. Forster Reaction

The Forster reaction involves the formation of secondary amines by the condensation of a primary amine with an aldehyde, followed by the addition of an alkyl halide to the Schiff base and subsequent hydrolysis.

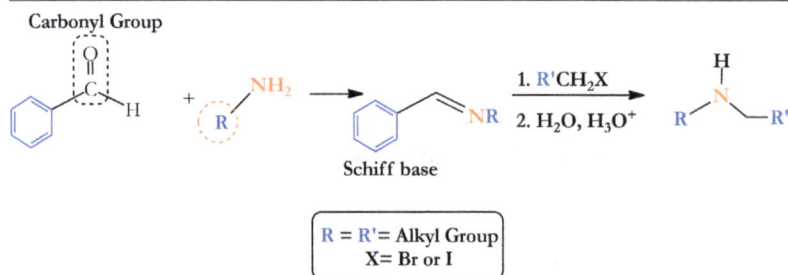

3. Baylis-Hillman Reaction

The Baylis-Hillman reaction is an organic reaction used to form a C–C bond between an α-β unsaturated carbonyl compound and an aldehyde. This reaction is most commonly catalyzed by DABCO (1,4-diazabicyclo[2.2.2]octane).

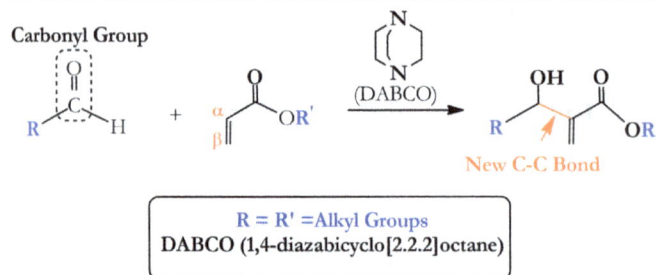

Table 9.1 (continued)

4. Dieckmann Condensation

Dieckmann condensation is an intramolecular chemical reaction of a diester with a base to give β-ketoester.

Carbonyl Groups

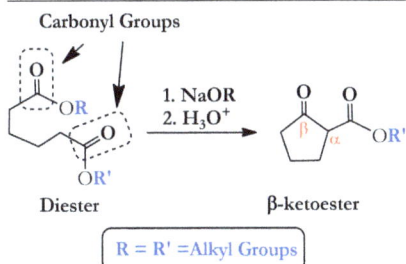

R = R' =Alkyl Groups

5. Clemmensen Reduction

In Clemmensen reduction aldehydes and ketones are converted to hydrocarbons (C=O to CH_2) after treatment with zinc amalgam, Zn(Hg), in concentrated hydrochloric acid.

Carbonyl Group

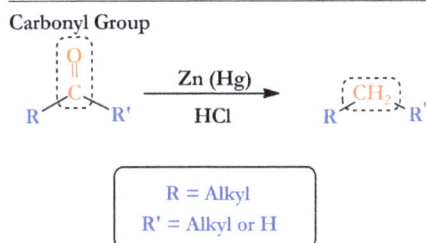

R = Alkyl
R' = Alkyl or H

6. Wolff-Kishner Reduction

In the Wolff-Kishner reduction, aldehydes and ketones are converted to hydrocarbons (C=O to CH_2) after being treated with hydrazine and heated with potassium hydroxide in ethylene glycol.

Carbonyl Group

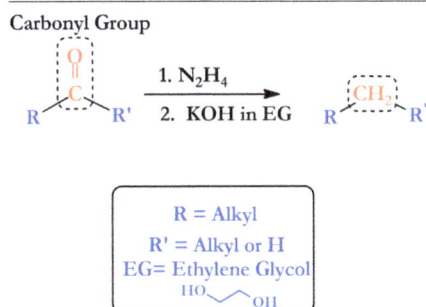

R = Alkyl
R' = Alkyl or H
EG= Ethylene Glycol

Table 9.1 (continued)

7. Emmert Reaction

The Emmert reaction involves the condensation of pyridine (or its analogs) with aldehydes or ketones in the presence of aluminum or magnesium amalgam to form pyridyl alkyl or pyridyl dialkyl carbinols.

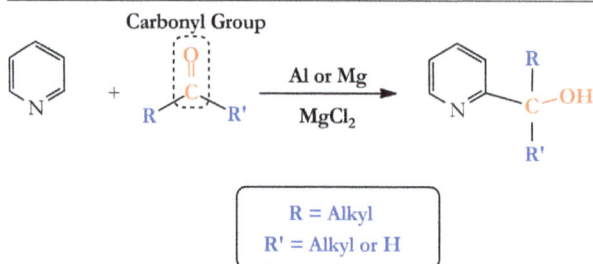

> R = Alkyl
> R' = Alkyl or H

8. Claisen Condensation

Esters containing both α-hydrogens and a carbonyl functional group can undergo a reversible condensation similar to the aldol reaction called a Claisen condensation. In which one ester acts as a nucleophile while a second ester acts as an electrophile.

During the reaction, a new carbon-carbon bond is formed to produce a β-keto ester product.

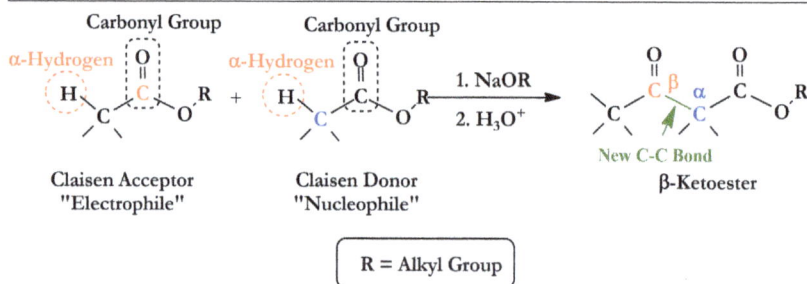

> R = Alkyl Group

9. Vilsmeier-Haack Reaction

The Vilsmeier-Haack reaction is a chemical reaction that produces an aryl aldehyde by reacting a substituted formamide such as dimethylformamide with phosphorus oxychloride ($POCl_3$) and an electron-rich aromatic ring.

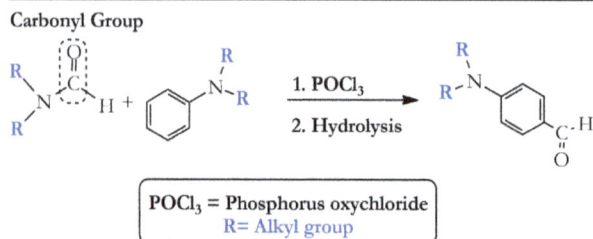

> $POCl_3$ = Phosphorus oxychloride
> R= Alkyl group

Table 9.1 (continued)

---

10. Thiele Reaction

In the thiele reaction, *p*-benzoquinone reacts with acetic anhydride in the presence of a catalytic amount of sulfuric acid or boron trifluoride (BF$_3$) to create 1,2,4-triacetoxybenzene.

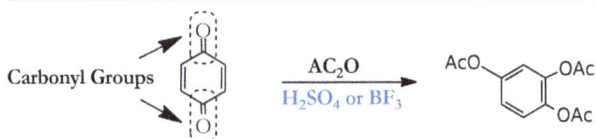

AC_2O = Acetic anhydride
BF_3 = Boron trifluoride
H_2SO_4 = Sulfuric acid

Table 9.2: Selected reactions that involve carbon-carbon double and triple bonds.

---

1. Simmons-Smith Reaction

The Simmons-Smith reaction is an organic cheletropic reaction involving an organozinc carbenoid that reacts with an alkene (or alkyne) to form a cyclopropane. This reaction affords the cyclopropanation of olefins.

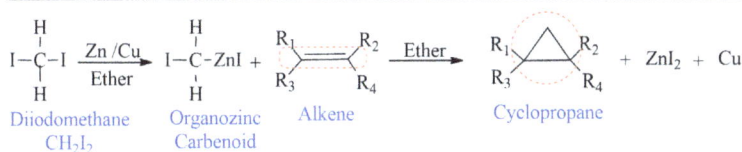

Diiodomethane CH_2I_2 Organozinc Carbenoid   Alkene   Cyclopropane

---

2. Ozonolysis

Ozonolysis is a weak oxidative cleavage where alkenes (double bonds) are cleaved into either ketones, aldehydes, or carboxylic acid using ozone.

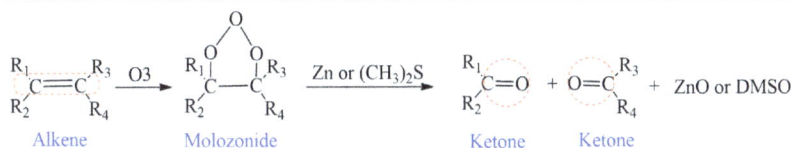

Alkene   Molozonide   Ketone   Ketone

---

3. Alkyne Zipper Reaction

Alkyne zipper reaction involves the isomerization of non-terminal alkynes to terminal alkynes. The base used in this reaction is potassium 1,3-diaminopropane, generated in situ by adding potassium hydride to the solvent 1,3-diaminopropane.

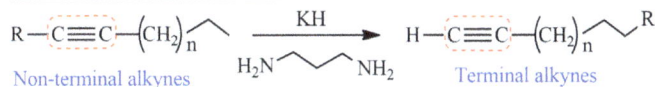

Non-terminal alkynes   Terminal alkynes

Table 9.2 (continued)

4. Bergman Cyclization

Bergman cyclization involves the thermal or photochemical cycloaromatization of enediyne in the presence of a hydrogen donor such as 1,4-cyclohexadiene to produce substituted arenes.

5. Brown Hydroboration

The syn-addition of hydroboranes to alkenes is known as Brown hydroboration. It is a two-step hydration process that transforms an alkene into an alcohol. This anti-Markovnikov reaction of hydroboration-oxidation results in the syn-addition of hydrogen to more-substituted carbon and a hydroxyl group to the less-substituted carbon.

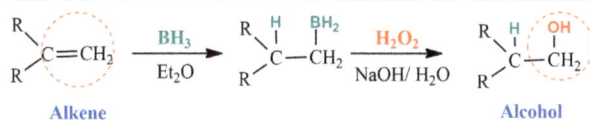

6. Cope Rearrangement

The Cope rearrangement deals with the rearrangement (thermal isomerization) of 1,5-dienes and related compounds in a reversible process.

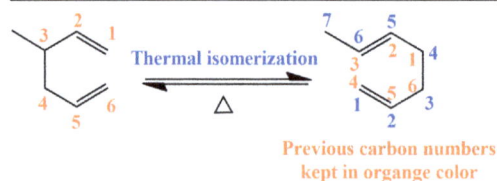

7. Glaser-Eglinton-Hay Coupling

This coupling reaction involves the synthesis of symmetric diacetylenes via the coupling of terminal alkynes through the reaction of the ethylenic compound with cuprous salts in the presence of oxygen and a base that is usually tetramethylethylenediamine (TMEDA or TEMED).

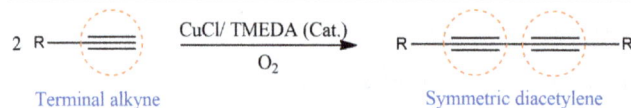

Table 9.2 (continued)

---

8. Kucherov Reaction

Alkynes react with water in the presence of dilute sulfuric acid (H_2SO_4) and mercury sulfate ($HgSO_4$) as a catalyst to produce aldehydes or ketones.

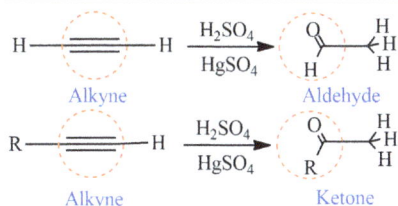

Alkyne → Aldehyde

Alkyne → Ketone

---

9. Oxymercuration-Reduction

In oxymercuration-demercuration reaction, the alkene is treated with mercury (II) acetate ($Hg(OAc)_2$) to produce an organomercury product, which is then treated with a reducing agent such as sodium borohydride ($NaBH_4$) to produce an alcohol, a Markovnikov addition product in which the OH group bonds to the more substituted carbon atom of the alkene.

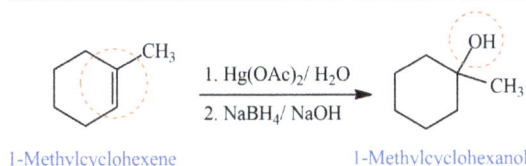

1. $Hg(OAc)_2$/ H_2O
2. $NaBH_4$/ NaOH

1-Methylcyclohexene → 1-Methylcyclohexanol

---

10. Upjohn Dihydroxylation

By using osmium tetroxide (OsO_4) as a catalyst and a stoichiometric quantity of an oxidant such as NMO (N-methyl morpholine-N-oxide), the Upjohn dihydroxylation allows for the syn-selective production of 1,2-diols from alkenes.

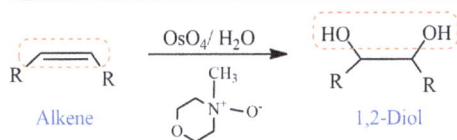

Alkene → 1,2-Diol

9.3 Organic Reactions Involve Alkyl Halides and Aryl Halides

Alkyl halide and aryl halide reactions are very important in the creation of desired chemical compounds that have been and continue to be useful in a variety of applications.

Among these reactions are the Suzuki-Miyaura reaction, the Arbuzov reaction, the Ullmann reaction, the Buchwald-Hartwig reaction, the Wurtz-Fittig reaction, the Fittig reaction, the Swarts reaction, the Heck reaction, the Hiyama Coupling reaction, and the Williamson reaction. Table 9.3 summarizes all of these reactions.

Table 9.3: Selected reactions that involve alkyl and aryl halides.

---

1. Suzuki-Miyaura Reaction

Suzuki-Miyaura reaction "Suzuki coupling" is based on the palladium (Pd) metal-catalyzed coupling of organohalide such as an aryl or vinyl (alkenyl) halide with an aryl or vinyl (alkenyl), or alkynyl organoborane (boronic acid or boronic ester, or in special cases with aryltrifluoroborane) under basic conditions. This coupling reaction is widely used to synthesize polyolefins, styrenes, and substituted biphenyls.

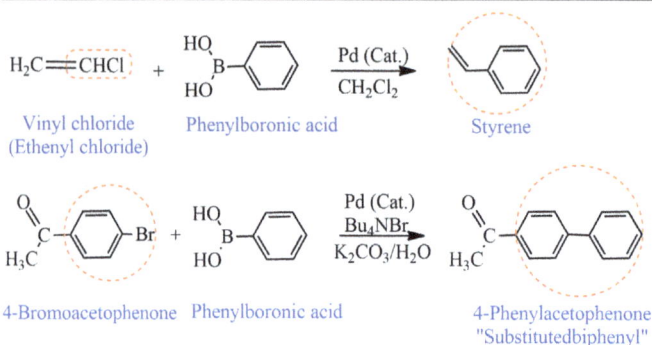

Vinyl chloride   Phenylboronic acid   Styrene
(Ethenyl chloride)

4-Bromoacetophenone Phenylboronic acid   4-Phenylacetophenone
                                                "Substitutedbiphenyl"

---

2. Arbuzov Reaction

The Michaelis-Arbuzov reaction "Arbuzov reaction" is a chemical reaction that occurs when a trivalent phosphorus ester reacts with an alkyl halide to generate a pentavalent phosphorus species and another alkyl halide.

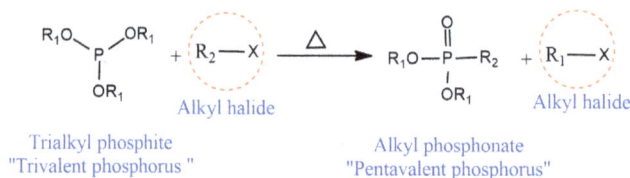

Trialkyl phosphite   Alkyl phosphonate
"Trivalent phosphorus "   "Pentavalent phosphorus"

---

3. Ullmann Reaction

The Ullmann reaction "Ullmann coupling" is an organic reaction that involves the coupling of two aryl halides in the presence of copper to yield a biaryl product.

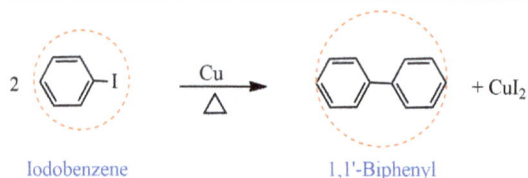

Iodobenzene   1,1'-Biphenyl

Table 9.3 (continued)

---

4. Buchwald-Hartwig Reaction
Buchwald-Hartwig amination reaction is used to form carbon-nitrogen bonds via the palladium-catalyzed coupling reaction of amines with aryl halides.

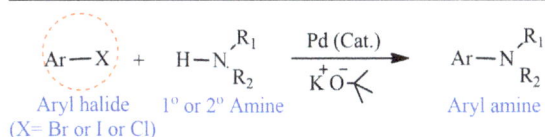

$$Ar-X \; + \; H-N\overset{R_1}{\underset{R_2}{}} \quad \xrightarrow[\overset{+}{K}\overset{-}{O}\text{-<}]{Pd\,(Cat.)} \quad Ar-N\overset{R_1}{\underset{R_2}{}}$$

Aryl halide   1° or 2° Amine   Aryl amine
(X= Br or I or Cl)

---

5. Wurtz-Fittig Reaction
In the Wurtz-Fittig reaction, aryl halides combine with alkyl halides and sodium metal in the presence of dry ether to produce substituted aromatic compounds such as toluene.

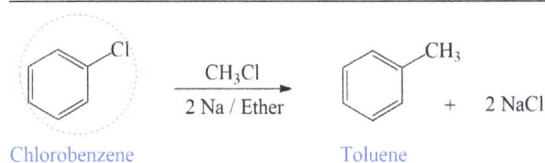

$$\text{Chlorobenzene} \xrightarrow[\text{2 Na / Ether}]{CH_3Cl} \text{Toluene} \; + \; 2\,NaCl$$

Chlorobenzene   Toluene

---

6. Fittig Reaction
Fittig reaction is a coupling reaction in which two aryl halides combine in the presence of sodium in dry ether or tetrahydrofuran to form a biaryl (diphenyl) species.

$$2\;\text{(Halobenzene)}-X \xrightarrow[\text{Dry ether}]{2\,Na} \text{(1,1'-Biphenyl)} \; + \; 2\,NaX$$

Halobenzene   1,1'-Biphenyl

---

7. Swarts Reaction
Swarts reaction is commonly used to synthesize alkyl fluorides from alkyl chlorides or alkyl bromides. This is achieved by heating the alkyl chloride or alkyl bromide in the presence of heavy metal fluorides such as silver fluoride or mercurous fluoride.

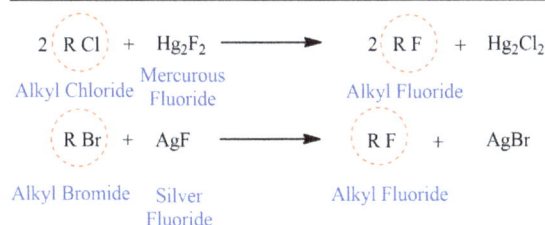

$$2\;R\,Cl \; + \; Hg_2F_2 \longrightarrow 2\;R\,F \; + \; Hg_2Cl_2$$

Alkyl Chloride   Mercurous Fluoride   Alkyl Fluoride

$$R\,Br \; + \; AgF \longrightarrow R\,F \; + \; AgBr$$

Alkyl Bromide   Silver Fluoride   Alkyl Fluoride

Table 9.3 (continued)

8. Heck Reaction

Palladium-catalyzed carbon-carbon coupling between aryl halides or vinyl halides and activated alkenes in the presence of a base to form more substituted alkene.

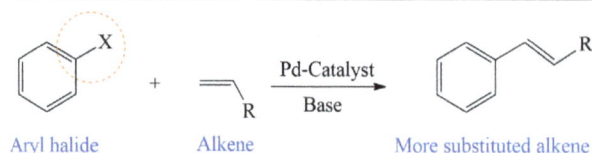

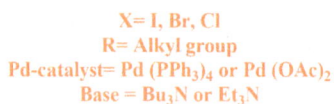

Aryl halide   Alkene   More substituted alkene

$$X = I, Br, Cl$$
$$R = Alkyl\ group$$
$$Pd\text{-}catalyst = Pd\ (PPh_3)_4\ or\ Pd\ (OAc)_2$$
$$Base = Bu_3N\ or\ Et_3N$$

9. Hiyama Coupling Reaction

This is a palladium-catalyzed cross-coupling reaction of organosilanes with alkyl or aryl halides to form carbon-carbon bonds.

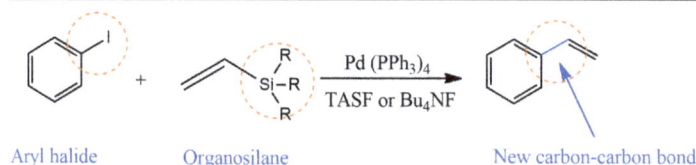

Aryl halide   Organosilane   New carbon-carbon bond

10. Williamson Reaction

In this reaction, an alkyl halide and an alkoxide or phenoxide react to generate an ether.

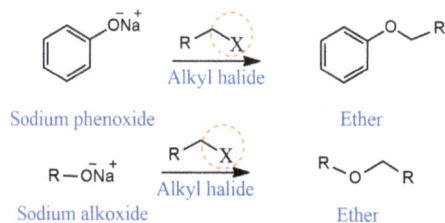

Sodium phenoxide   Ether

Sodium alkoxide   Ether

$$R = Alkyl\ group$$
$$X = I, Br, Cl$$

9.4 Organic Reactions Involve Alcohols

Alcohol-related reactions are critical in the formation of desired chemical compounds that have been and continue to be beneficial in a range of applications. Among these reactions are the Mitsunobu reaction, Jones oxidation reaction, Ley-Griffith reaction, Swern oxidation, Criegee oxidation, Oppenauer oxidation, Pinacol-Pinacolone rearrangement, Haloform reaction, and Fischer esterification. Table 9.4 summarizes all of these reactions.

Table 9.4: Selected reactions that involve alcohols.

---

1. Mitsunobu Reaction

The Mitsunobu reaction is a chemical reaction that uses an azodicarboxylate such as diethyl azodicarboxylate (DEAD) or diisopropyl azodicarboxylate (DIAD) and triphenylphosphine to convert primary or secondary alcohol into thioethers, phenyl ethers, esters, and numerous other compounds.

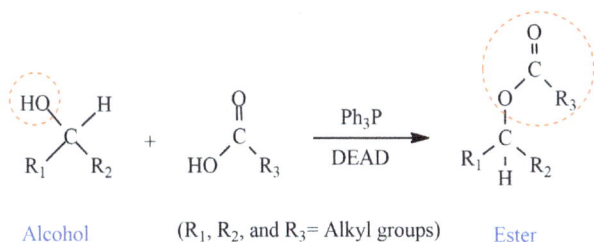

Alcohol   (R_1, R_2, and R_3= Alkyl groups)   Ester

---

2. Jones Oxidation

The Jones oxidation is a chemical reaction in which primary and secondary alcohols are oxidized to carboxylic acids and ketones, respectively. Jones Reagent is aqueous acetone solution of CrO_3. Because it is a gentle reagent, it oxidizes alcohols without oxidizing or rearranging double bonds.

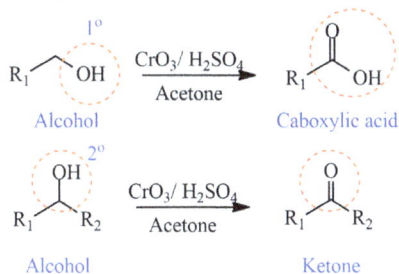

Alcohol   Caboxylic acid

Alcohol   Ketone

---

3. Ley-Griffith Reaction

The oxidation of primary or secondary alcohols to aldehydes or ketones is known as the Ley-Griffith reaction. The primary catalyst is tetrapropylammonium perruthenate (TPAP), which is combined with the co-oxidant *N*-methylmorpholine *N*-oxide (NMO).

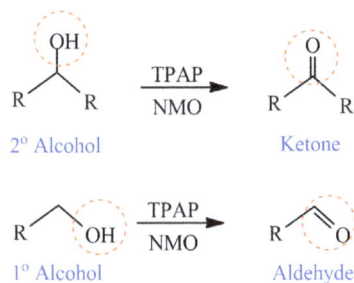

2° Alcohol   Ketone

1° Alcohol   Aldehyde

Table 9.4 (continued)

4. Alcohol-Based Michaelis-Arbuzov Reaction

A wide range of alcohols can readily react with phosphites, phosphonites, and phosphinites to give all three kinds of phosphoryl compounds (phosphonates, phosphinates, and phosphine oxides) using an n-Bu$_4$NI-catalyzed efficient C–P(O) bond formation reaction.

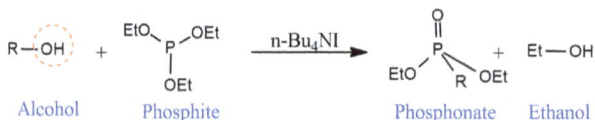

R—OH + [EtO–P(OEt)–OEt] $\xrightarrow{n\text{-Bu}_4\text{NI}}$ [EtO–P(O)(R)–OEt] + Et—OH

| Alcohol | Phosphite | | Phosphonate | Ethanol |

5. Swern oxidation

A chemical reaction that converts a primary or secondary alcohol (OH) to an aldehyde (CHO) or ketone (–C=O) using oxalyl chloride, dimethyl sulfoxide (DMSO), and an organic base such as triethylamine (Et$_3$N).

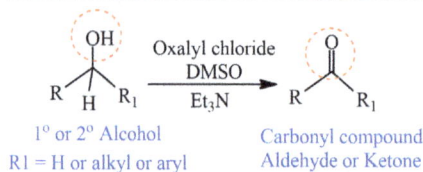

(R)(H)(R$_1$)C–OH $\xrightarrow[\text{Et}_3\text{N}]{\substack{\text{Oxalyl chloride}\\ \text{DMSO}}}$ R–C(=O)–R$_1$

1° or 2° Alcohol   Carbonyl compound
R1 = H or alkyl or aryl   Aldehyde or Ketone

6. Criegee Oxidation

This reaction involves the use of lead tetraacetate [Pb (OAc)$_4$] to oxidize vicinal diols to ketones and aldehydes. This glycol cleavage reaction can also be achieved by several oxidizing agents including periodic acid (HIO$_4$), PIDA [PhI (OAc)$_2$], cerium (IV) salts, and silver(I) salts.

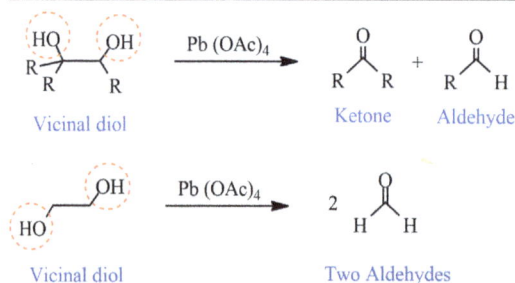

R–C(HO)(R)–C(OH)(R) $\xrightarrow{\text{Pb (OAc)}_4}$ R–C(=O)–R + R–C(=O)–H

Vicinal diol   Ketone   Aldehyde

HO–CH$_2$–CH$_2$–OH $\xrightarrow{\text{Pb (OAc)}_4}$ 2 H–C(=O)–H

Vicinal diol   Two Aldehydes

Table 9.4 (continued)

7. Oppenauer Oxidation

The aluminum-catalyzed hydride shift from the -carbon of the alcohol component to the carbonyl carbon of a second component is known as Meerwein-Ponndorf-Verley-Reduction (MPV) or Oppenauer Oxidation (OPP) depending on the isolated product. Oppenauer Oxidation is used when the desired products are aldehydes or ketones. As a catalyst, aluminum isopropoxide Al(O-i-Pr)$_3$ is commonly used.

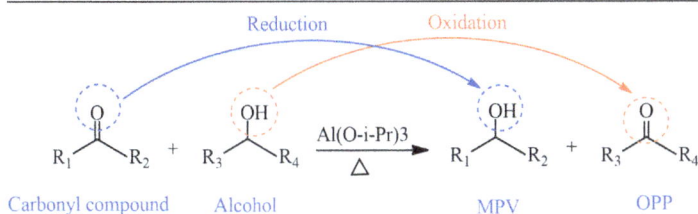

8. Pinacol-Pinacolone Rearrangement

Pinacol-pinacolone rearrangement involves the conversion of 1,2-diols into carbonyl compounds. The 1,2-rearrangement takes place under acidic conditions.

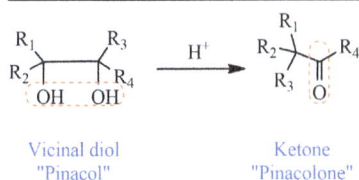

Vicinal diol "Pinacol"   Ketone "Pinacolone"

9. Haloform Reaction

When a methyl ketone is exposed to a base and a halogen such as I$_2$, Br$_2$, or Cl$_2$, it is converted into a carboxylic acid and a haloform (HCX$_3$).

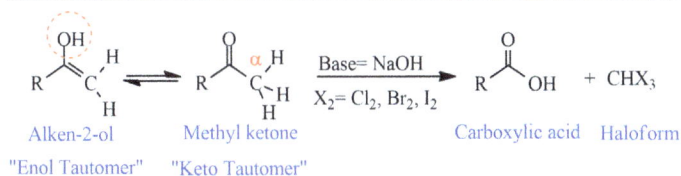

Alken-2-ol   Methyl ketone   Carboxylic acid   Haloform
"Enol Tautomer"   "Keto Tautomer"

10. Fischer Esterification or Fischer-Speier Esterification

It is a form of esterification that occurs when a carboxylic acid and an alcohol are refluxed in the presence of an acid catalyst.

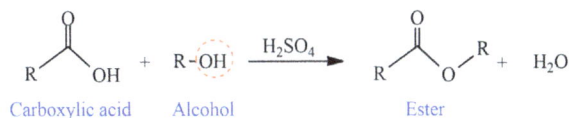

Carboxylic acid   Alcohol   Ester

9.5 Essential Terms

Pinacol A diol that has hydroxyl groups on vicinal carbon atoms.

Pinacolone An organic compound containing ketone functional group.

Esterification A reaction of combining organic acid with alcohol to form ester.

Haloform A compound derived from methane by substitution of three hydrogens by halogen atoms. Chloroform is an example.

Vicinal diol Two hydroxyl groups occupy vicinal positions by attaching to adjacent atoms.

Phosphite A salt of phosphorous acid ester. This trivalent inorganic anion is obtained by the removal of all three hydrogens from the phosphorous acid. Phosphite's general formula is $(RO)_3P$.

Phosphonite An organophosphorus compound with a general formula $(RO)_2 (R)P$.

Phosphinite An organophosphorus compound with a general formula $(R)_2 (OR)P$.

Phosphonate An organophosphorus compound with a general formula $(R) (OR)_2P=O$.

Phosphinate An organophosphorus compound with a general formula $(R)_2 (OR)P=O$.

Phosphine oxide An organophosphorus compound with a general formula $(R)_3 P=O$ or $(Ar)_3 P=O$. Triphenyl phosphine oxide is an example.

Coupling reaction A type of organic reaction in which two molecules react together to form a new one. An example is the formation of dipeptide from two molecules of amino acids.

Dihydroxylation A reaction by which alkene is converted into diol. An example is the addition of two hydroxyl groups to the double bond.

Condensation reaction A reaction in which two molecules are combined to form one molecule after releasing small molecule such as water. An example is the esterification reaction.

9.6 Problems

9.6.1   Draw the chemical structures of NMO and TPAP.

9.6.2   Write the full name of the following catalyst **n-Bu₄NI**.

9.6.3   Draw the chemical structure of aluminum isopropoxide catalyst.

9.6.4   What is a pinacol? Give an example.

9.6.5   What is pinacol pinacolone rearrangement?

9.6.6   Describe the pinacol pinacolone rearrangement mechanism.

9.6.7   What is the difference between enolizable and non-enolizable alpha carbon?

9.6.8   Show how methyl benzoate is formed by Fischer esterification.

9.6.9   What kind of halides are used in the Nozaki-Hiyama-Kishi reaction?

9.6.10 Draw the chemical structure of Schiff base.

9.6.11 What is the full name of DABCO catalyst?

9.6.12 Show by structure the final product of the Dieckmann condensation.

9.6.13 What is the zinc amalgam?

9.6.14 Describe the Clemmensen reduction mechanism.

9.6.15 What kind of esters are involved in the Claisen condensation?

9.6.16 Draw the chemical structure of the organozinc carbenoid.

9.6.17 Name the reaction that involves the isomerization of non-terminal alkynes to terminal alkynes.

9.6.18 Give an example of organic hydrogen donor, draw its structure, and mention its use.

9.6.19 Define Cope rearrangement.

9.6.20 What kind of base is used in the Glaser-Eglinton-Hay coupling?

9.6.21 Describe the oxymercuration-demercuration reaction mechanism.

9.6.22 What kind of organoboranes is involved in the Suzuki-Miyaura reaction?

9.6.23 Give one example for trivalent phosphorus and another for pentavalent phosphorus.

9.6.24 Which catalyst is used in the Buchwald-Hartwig amination reaction?

9.6.25 Which reaction can be used to make biphenyl compounds?

9.6.26 What kind of nucleophiles is used in the Williamson reaction?

9.6.27 What do DEAD and DIAD stand for?

9.6.28 What is Jones reagent and what it is used for?

9.6.29 What is the primary catalyst in the Ley-Griffith reaction?

9.6.30 What kind of oxidizing agent can be used to facilitate the cleavage of glycol?

∗∗∗∗∗∗∗∗∗∗∗

Solutions to Problems

Chapter 1

1.9.1

Carboxyl group

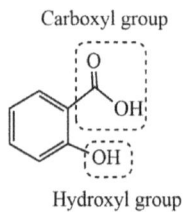

Hydroxyl group

1.9.2.

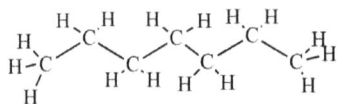   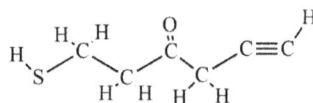

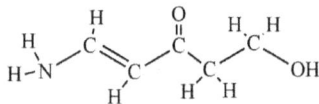

1.9.3 $C_4H_{10}O$

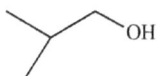

C_4H_8O

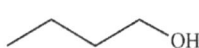

https://doi.org/10.1515/9783111382753-010

$C_4H_8O_2$

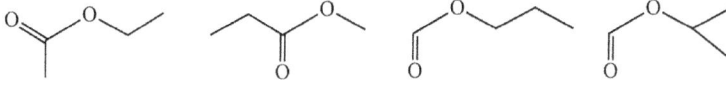

1.9.4 Carbon = 4
       Sulfur = 2
       Nitrogen =3
       Oxygen = 2

1.9.5

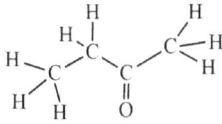

1.9.6

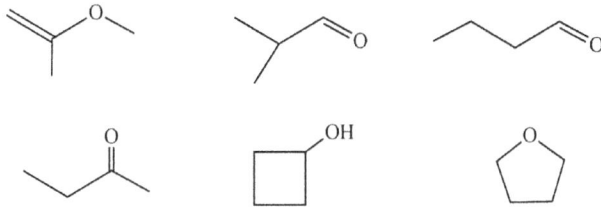

1.9.7

Family Name	Functional Group Structure
Isocyanates	R-N=C=O
Nitriles	R-CN
Sulfoxides	R-SO-R

1.9.8

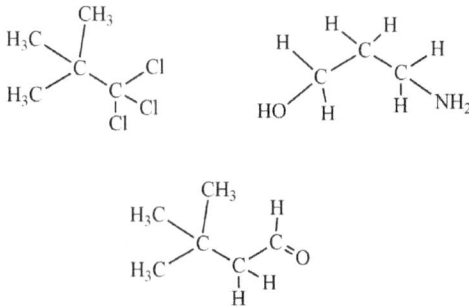

1.9.9

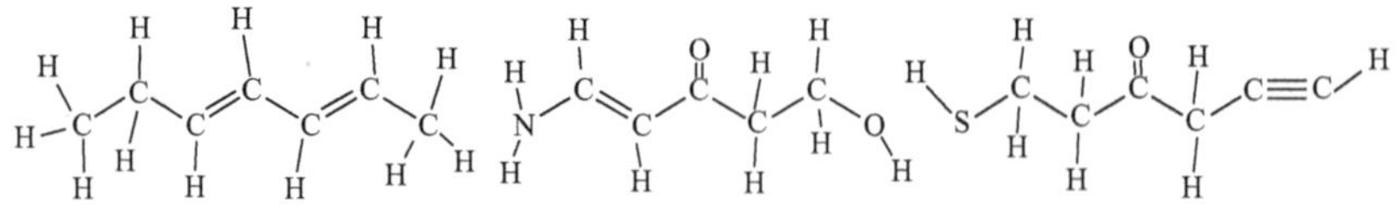

1.9.10

Functional group structure	Family name
R-COOH	Carboxylic acids
R-N=N-R	Azo-compounds
$RCONH_2$	Amides

1.9.11 Nine carbon atoms

1.9.12 Fourteen hydrogen atoms

1.9.13 CH_4 and NH_3

1.9.14

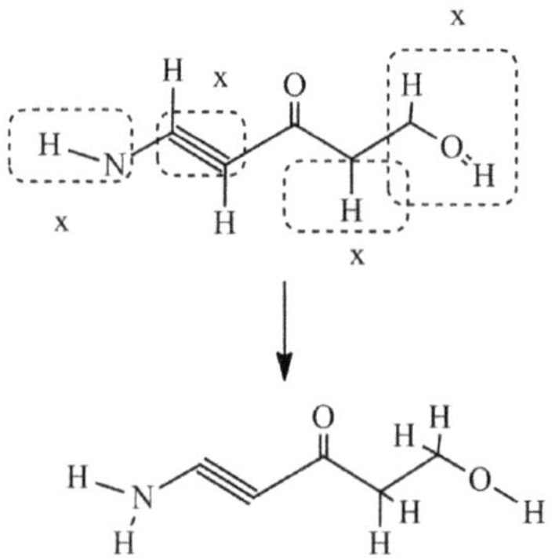

Correct structure

1.9.15

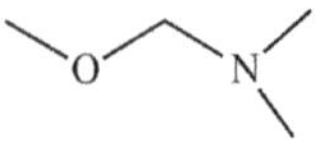

$CH_3OCH_2N(CH_3)_2$

1.9.16 Four carbon-carbon σ bonds

1.9.17 Five π bonds

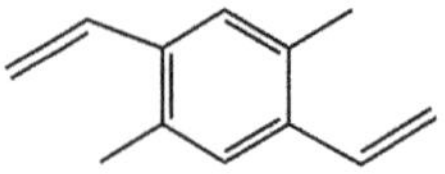

1.9.18

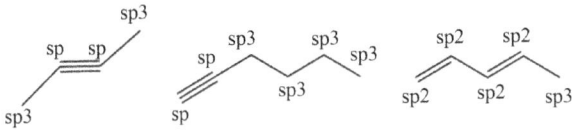

1.9.19

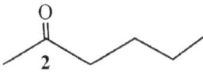

1.9.20 Both

1.9.21

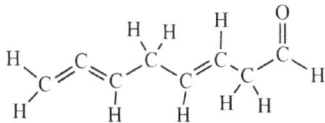

1.9.22 $CH_3COCH_2CH_2CH_2CH_3$
$CH_3CH_2COCH_2CH_2CH_3$
$CH_3COCH(CH_3)CH_2CH_3$
$CH_3COCH_2CH(CH_3)CH_3$
$CH_3CH_2COCH(CH_3)CH_3$

1.9.23

Pure atomic orbitals of the central atom	Hybridization of the central atom
s, p, p	sp2
s, p	sp

1.9.24 Two sp^2 and three sp

1.9.25 One S and two P

1.9.26

$CH_2=C=CHCH_2CH=CHCH_2CHO$

1.9.27

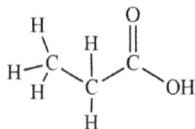

1.9.28

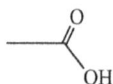

1.9.29 The correct one is CCl_4.
The correct one is H_2S.
The correct one is N_3H_3.

1.9.30

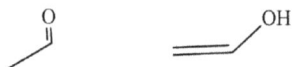

****

Chapter 2

2.11.1

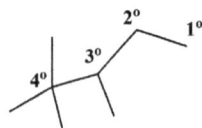

2.11.2

2.11.3

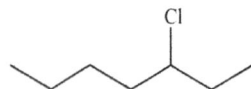

2.11.4 C_nH_{2n} an example cyclobutane C_4H_8

2.11.5

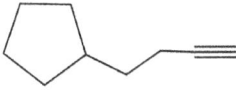

Nine carbons and fourteen hydrogens

2.11.6 Five methylene groups (-CH$_2$)

2.11.7 2-Methylbutane

2.11.8

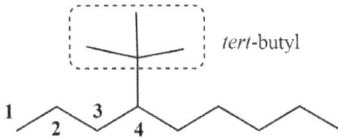

2.11.9 Four methyl groups (-CH$_3$)

2.11.10 3-*tert*-butyl-2,2-dimehtylbutane

2.11.11

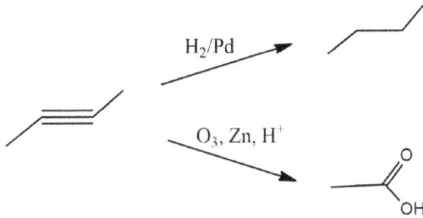

2.11.12

No.	Reactions
1	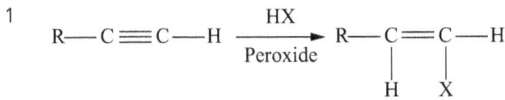
2	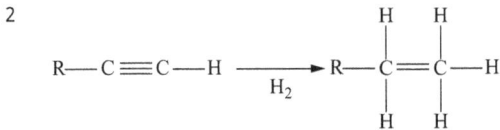
3√	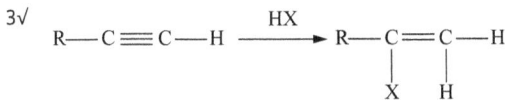

2.11.13

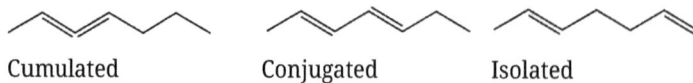

Cumulated   Conjugated   Isolated

2.11.14

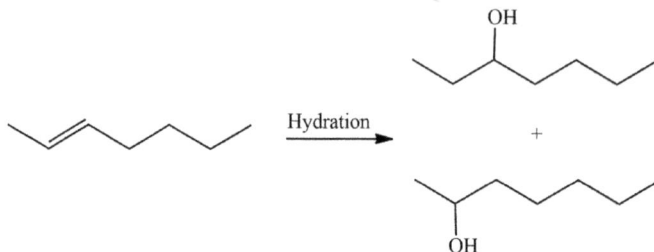

Hydration

2.11.15 $CnH_{2(n-m)}$, an example is cyclohexene, C_6H_{10}

2.11.16

IUPAC name	Structure
5-Methyl-2-octyne	
4-Chloro-2-pentene	
1,3-Dibromocyclohexane	

2.11.17 $(CH_3)_2CHCH_2CH_2CH=CH_2$
5-Methyl-2-hexene

2.11.18

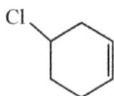

4-Chlorocyclohex-1-ene

2.11.19

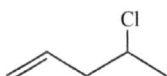

4-Chloropent-1-ene

2.11.20

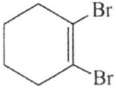

1,2-Dibromocyclohexene

2.11.21

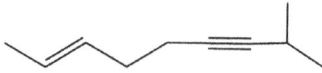

8-Methyl-2-nonen-6-yne

2.11.22

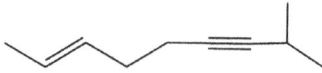

Terminal Alkyne   **Internal Alkyne**

2.11.23 5% Pd-CaCO$_3$, Pb (OCOCH$_3$)$_2$, quinoline

OR

Pd-BaSO$_4$, quinoline, methanol

2.11.24

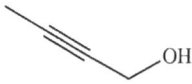

2.11.25

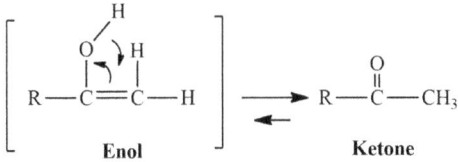

Enol   Ketone

2.11.26

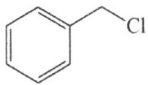

The halogen atom (Cl) is not attached directly to the aromatic ring

2.11.27

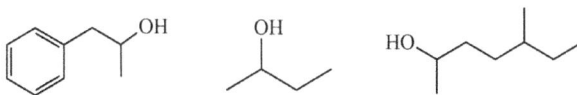

2.11.28

2.11.29

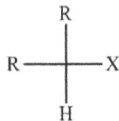

2.11.30

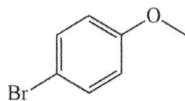

Secondary alkyl halide

****

Chapter 3

3.9.1

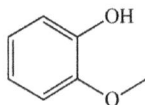

p-Bromoanisole

2-Methoxyphenol

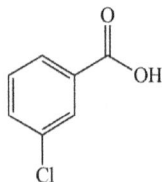

m-Chlorobenzoic acid

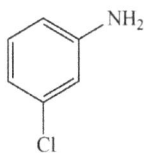

3-Chloroaniline

o-Ethyltoluene

3.9.2   The electrophile is Cl$^+$ (chloronium ion)

3.9.3

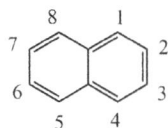

3.9.4

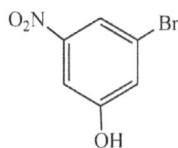

3-Bromo-5-nitrophenol

3.9.5   *m*-Nitrophenol

3.9.6

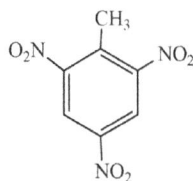

2,4,6-Trinitrotoluene

3.9.7

   i.   Phenol (Hydroxybenzene)

   ii.   Toluene (Methylbenzene)

   iii.   Aniline (Aminobenzene)

3.9.8

   i.   Cl$^+$ (chloronium ion)

   ii.   R$^+$ (carbocation)

   iii.   RCO$_+$ (acylium ion)

3.9.9 Used as a basic component of polystyrene and styrofoam plastics

3.9.10 The $-CH_3$ substituent is ortho- and para-directing and activator

3.9.11 *m*-Dihydroxybenzene

3.9.12

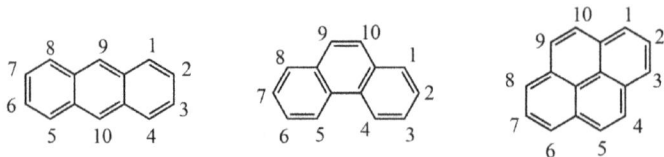

3.9.13

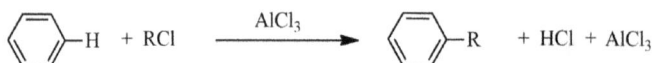

3.9.14 *p*-Phenolsulfonic acid

3.9.15 Chloro derivatives are used extensively as insecticides, herbicides, fungicides, and bactericides

3.9.16

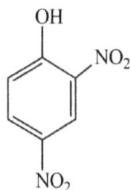

3.9.17

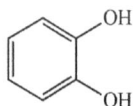

3.9.18 Isopropylbenzene

3.9.19

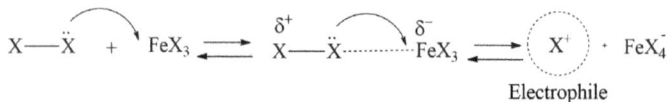

Assist in the generation of the electrophile

3.9.20 Electrophilic aromatic substitution

3.9.21 R-X

3.9.22

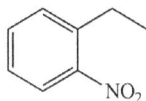

3.9.23 Because aromatic nucleophilic substitution is preferable for aromatic compounds and haloalkanes are aliphatic compounds.

3.9.24 RMgX

An example is CH_3MgBr

3.9.25

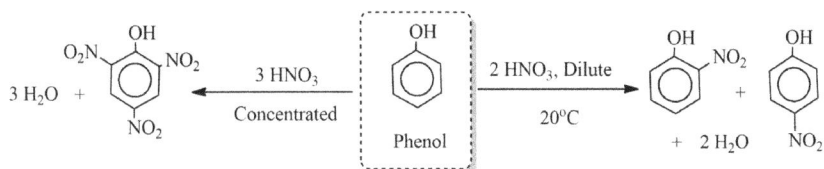

3.9.26

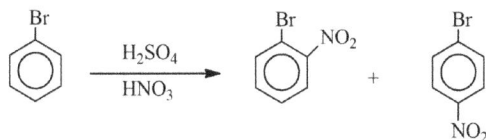

3.9.27

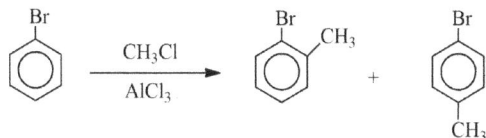

3.9.28

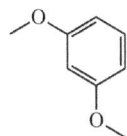

3.9.29 Catechol (1,2-hydroxybenzene)

Resorcinol (1,3-hydroxybenzene)

Hydroquinone (1,4-hydroxybenzene)

3.9.30

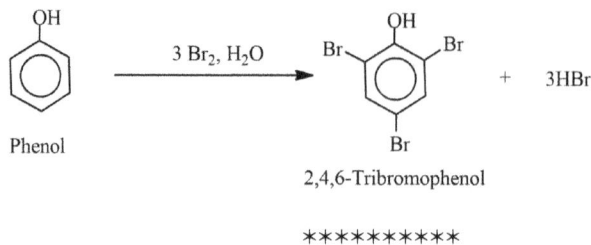

Phenol   2,4,6-Tribromophenol

★★★★★★★★★★

Chapter 4

4.8.1

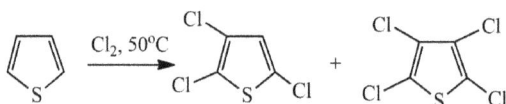

4.8.2

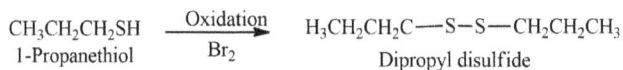

4.8.3

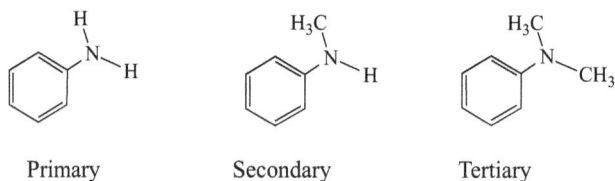

Primary   Secondary   Tertiary

4.8.4 Because they can act only as hydrogen bond acceptor

4.8.5 Diethyl sulfide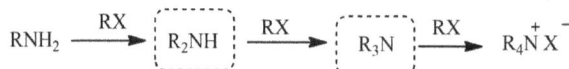

Ethyl methyl sulfide

2-Mercaptohexanol

1-Decanethiol

4.8.6

$$RNH_2 \xrightarrow{\text{RX}} R_2NH \xrightarrow{\text{RX}} R_3N \xrightarrow{\text{RX}} R_4\overset{+}{N}\,X^-$$

4.8.7 The most common ones are oxygen, nitrogen, and sulfur.

4.8.8

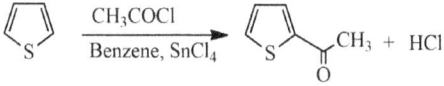

$$\text{CH}_3\text{COCl} \xrightarrow{\text{Benzene, SnCl}_4}$$

... CH_3 + HCl

4.8.9

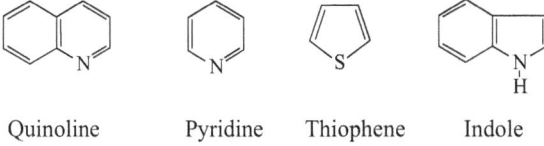

Quinoline   Pyridine   Thiophene   Indole

4.8.10

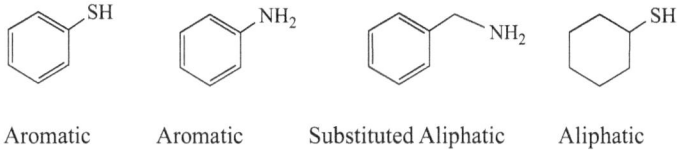

Aromatic   Aromatic   Substituted Aliphatic   Aliphatic

4.8.11 They are named as sulfides.

4.8.12

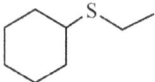

Cyclohexyl ethyl sulfide

4.8.13

4.8.14

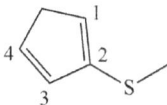

2-(Methylthio)cyclopent-1,3-diene

4.8.15

Azine

4.8.16

1,2-Oxazole

4.8.17 At carbon-5 and carbon-8

4.8.18

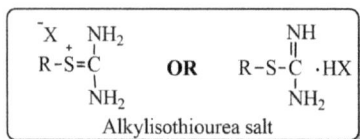

Alkylisothiourea salt

4.8.19

BME is β-mercaptoethanol

4.8.20 They are names as thiols or mercaptans

4.8.21

Diethyl sulfide

4.8.22

Tetrahydrofuran (THF)

4.8.23 18-Crown-6-ether stands for (18 as total atoms number and 6 for the number of oxygen atoms)

4.8.24

4.8.25

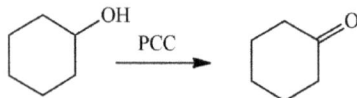

4.8.26 The product will be 2-hexanone

4.8.27 4,5-Dimethylheptan-2-ol is secondary alcohol
4.8.28 $(CH_3)_2CHCH_2OH$
          2-Methylpropanol
4.8.29 Hydrogen bonding
4.8.30

1-Oxa-2-thiacyclopenta-2,4-diene

∗∗∗∗∗∗∗∗∗∗

Chapter 5

5.6.1 $C_2H_5COC_2H_5$
        (Pentan-3-one)
        $C_2H_5COCH_3$
        (Butan-2-one)
        $CH_3(CH_2)_4CHO$
        (Hexanal)
        CH_3CH_2CHO
        (Propanal)
5.6.2

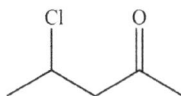

4-Chloro-2-pentanone

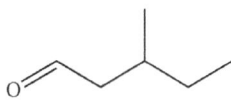

3-Methyl pentanal

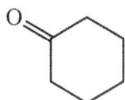

Cyclohexanone

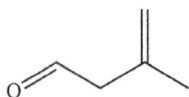

3-Methyl-3-butenal

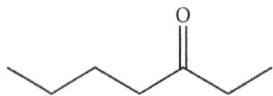

3-Heptanone

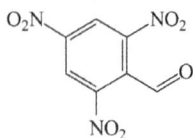

2,4,6-Trinitrobenzaldehyde

5.6.3 Hexanoic acid
       Octan-3-one
       Nonan-4-one
5.6.4 Benzaldehyde only
5.6.5 Toluene
5.6.6 $CH_3CH_2CH=CHCHO$
       (Pent-2-enal or 2-pentenal)
       $CH_3COCH_2CH_2CH_3$
       (2-Pentanone)
5.6.7

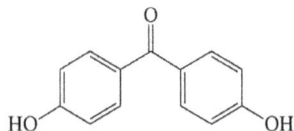

5.6.8

5.6.9   β-Hydroxy aldehyde (aldol) and β-hydroxy ketone (ketol)
5.6.10 Oxygen reacts with acids and electrophiles
5.6.11 Yes, because they can isomerize to (equilibrate with) aldoses via an enediol
        intermediate

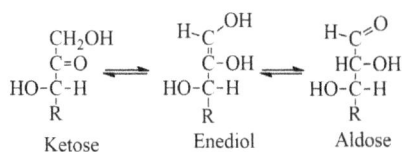

Ketose Enediol Aldose

5.6.12

$$\xrightarrow{\text{PCC, DCM}}$$

5.6.13 Tollen's test

5.6.14

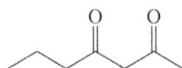

5.6.15

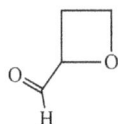

5.6.16

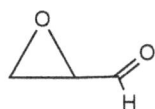

Oxirane-2-carbaldehyde

5.6.17 Liquid

5.6.18 Jones reagent is a solution of chromium trioxide in aqueous sulfuric acid

5.6.19 Pent-2-one

5.6.20 Aldehydes

5.6.21

   i.   Hexan-1-ol to hexanal (PCC or PDC)

   ii. Cyclopentanol to cyclopentanone (($COCl)_2$, DMSO, and TMA)

   iii. *p*-Chlorotoluene to *p*-chlorobenzaldehyde (CrO_2Cl_2)

   iv. Ethanenitrile to ethanal ($SnCl_2$ and HCl)

5.6.22

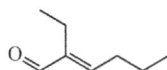

2-Ethylhex-2-enal

5.6. 23

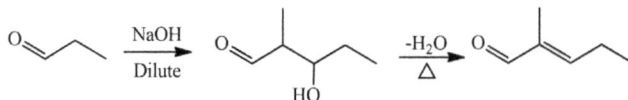

5.6.24 An aldehyde, such as benzaldehyde, methanal, and 2,2-dimethylpropanal, with no α-hydrogen atom undergoes self-oxidation and reduction reaction on heating with concentrated alkali. Reduction produces an alcohol while oxidation gives carboxylic acid salt

5.6.25 Aldol condensation occurs when aldehydes having α-hydrogen react with a dilute base to give β-hydroxy aldehydes called aldols. When the condensation reaction involves two different carbonyl compounds then it is called cross aldol condensation

5.6.26 An example of β-hydroxy ketone (ketol) is 4-hydroxy-4-methylpentan-2-one

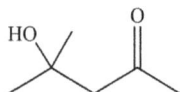

5.6.27 Only primary or secondary alcohols

5.6.28 The reaction in which benzenes and their derivatives are converted to benzaldehydes or substituted benzaldehydes by treatment with carbon monoxide and hydrogen chloride in the presence of anhydrous aluminum chloride or cuprous chloride

5.6.29

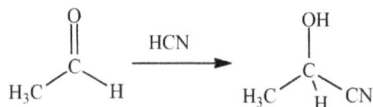

5.6.30 The Jones oxidation is the over-oxidation of primary alcohols to carboxylic acids and the oxidation of secondary alcohols to ketones.

✶✶✶✶✶✶✶✶✶✶

Chapter 6

6.5.1 Pentanoic acid (Valeric acid)
Hexanoic acid (Caproic acid)
Heptanoic acid (Enanthic acid)
Octanoic acid (Caprylic acid)

6.5.2

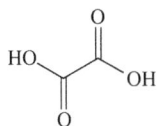

Oxalic acid

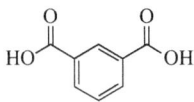

Isophthalic acid

6.5.3

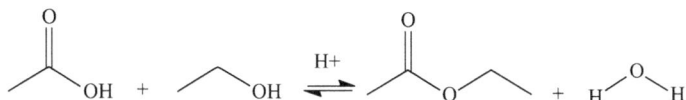

Ethanoic acid   Ethanol   Ethyl acetate   Water

6.5.4

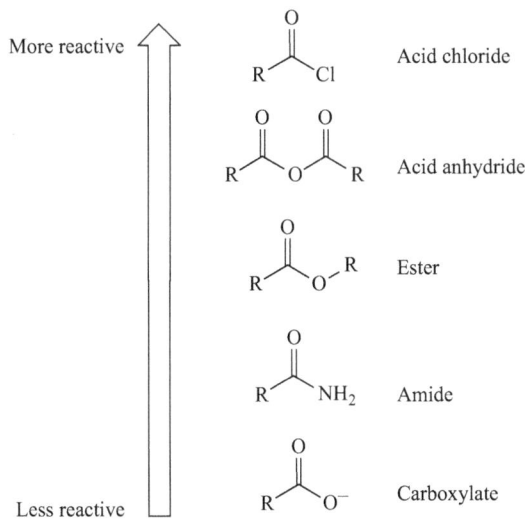

More reactive

Acid chloride

Acid anhydride

Ester

Amide

Carboxylate

Less reactive

6.5.5

Ester	Carboxylic acid
Methyl acetate	Acetic acid
Ethyl formate	Formic acid
Ethyl butyrate	Butyric acid
Methyl propionate	Propionic acid

6.5.6   Methyl ethanoate and methanoic acid

6.5.7

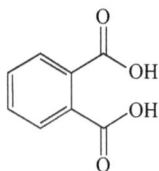

Phthalic acid   Isophthalic acid

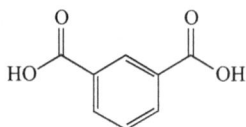

6.5.8

Ethanoic anhydride   Ethanoic benzoic anhydride
"Symmetrical anhydride"   "Mixed anhydride"

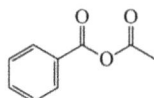

6.5.9   Because primary and secondary amides can act as hydrogen bond donor and acceptor, but tertiary amides can act as hydrogen bond acceptor only.

6.5.10

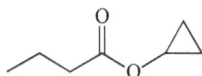

Cyclopropyl butanoate

6.5.11 $CH_3CH_2CH(CH_3)CH(NH_2)COOH$
2-Amino-3-methylpentanoic acid

6.5.12   The acid chloride

6.5.13   Methyl benzoate

6.5.14   Acetic acid

6.5.15

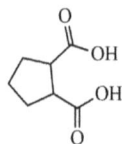

1,2-Cyclopentanedicarboxylic acid

6.5.16

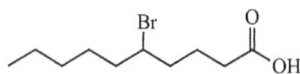

5-Bromodecanoic acid

6.5.17   The common name of hexanedioic acid is adipic acid.

6.5.18   Propanol

6.5.19

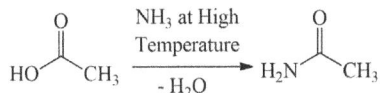

6.5.20 Lithium aluminum hydride (LiAlH$_4$)
6.5.21 Because they can form hydrogen bond with water
6.5.22 Methanol and ethanol
6.5.23

6.5.24 Acetamide
6.5.25 Malonic acid (used for making barbiturates)
Phthalic acid (used as plasticizer)
Glutaric acid (used in the production of polyols)
6.5.26 2-Phenyl-2-propanol
6.5.27

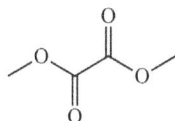

Dimethyl oxalate

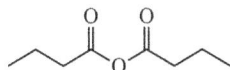

Butanoic anhydride

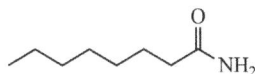

Octanamide

6.5.28 The common name of 2-amino-3-carbamoylpropanoic acid is asparagine.
6.5.29 The IUPAC name of the linoleic acid is Octadeca-9,12-dienoic acid.
6.5.30

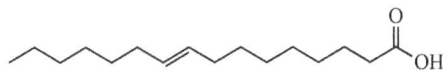

Hexadec-9-enoic acid

****

Chapter 7

7.9.1 Bifunctional monomers

7.9.2 Radical initiator

7.9.3 Substituents are arranged in the atactic polymers on random sides of the backbone.

7.9.4 Polymers are classified on molecular forces basis into elastomers, thermoplastics, thermosetting, and fibers.

7.9.5 In isotactic polymers substituents are positioned on the same side of the backbone.

7.9.6

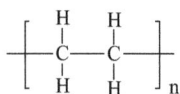

Polyethylene   Propylene

7.9.7 Phenol and formaldehyde

7.9.8 Addition polymers result from the addition of monomers to the growing chain. Condensation polymers are formed by the combination of two bifunctional monomers and releasing of small molecules.

7.9.9

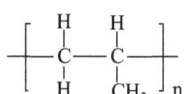

7.9.10

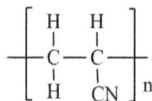

Poly(acrylonitrile)

7.9.11

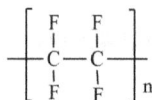

Polytetrafluoroethylene

7.9.12 Polystyrene

7.9.13 Atactic polymer

7.9.14

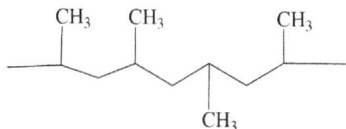

7.9.15

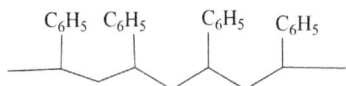

7.9.16 Sulfur

7.9.17 PVA is addition polymer.

7.9.18 Plasticizer

7.9.19

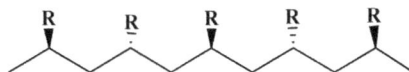

Syndiotactic

7.9.20 PMMA stands for poly(methyl methacrylate)

7.9.21 Stabilizers are chemical additives that protect polymers from environmental effects.

7.9.22 Linear, branched, and comb

7.9.23 43,000

7.9.24 Random – monomers A and B are positioned along chain randomly.
Alternating – monomers A and B are positioned in an alternate order within the chain.

7.9.25 $AlCl_3$ and $SnCl_4$

7.9.26 Styrene-butadiene

7.9.27 $CH_2=CHCl$
Vinyl chloride

7.9.28 Condensation

7.9.29 $C_6H_5-CH=CH_2$
Styrene

7.9.30

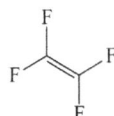

Tetrafluoroethylene (TFE)

****

Chapter 8

8.16.1   Cis isomers are compounds that have two comparable atoms on the same side of a molecule's double bond. Trans isomers are compounds that have two comparable atoms on opposing sides of a double bond.

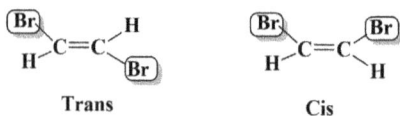

Trans   Cis

8.16.2 CH_3-OH, CH_3-CH_2-OH, $CH_3CH_2CH_2$-OH,
These are members of a homologous series called "Alcohols."

8.16.3   They are skeletal isomers which is type of structural (constitutional) isomerism.

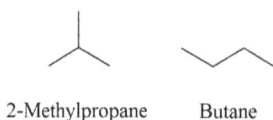

2-Methylpropane   Butane

8.16.4

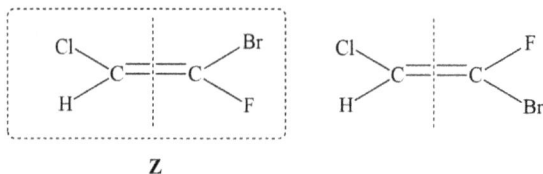

Z

Because the atoms of higher atomic number are on the same side of the double bond.

8.16.5

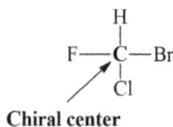

Chiral center

8.16.6

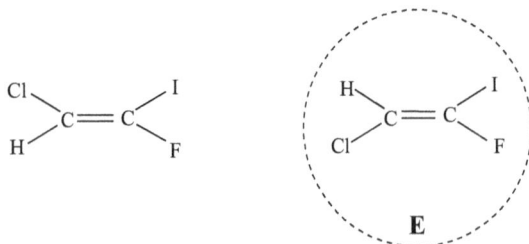

E

8.16.7
- Positional
- Functional
- Skeletal

8.16.8

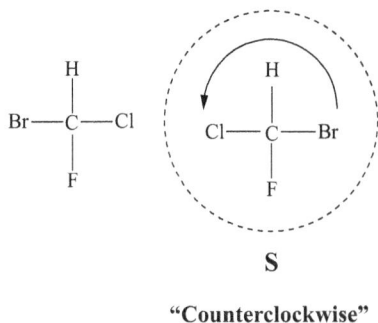

S

"Counterclockwise"

8.16.9

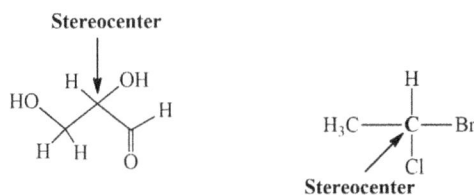

8.16.10 Yes, because it fulfills all the characteristics of homologous series.
CH_3-CHO, CH_3-CH_2-CHO, CH_3-CH_2-CH_2-CHO

8.16.11 A series of compounds, like the aldehydes, that differ only by the number of –CH_2– groups, is called a___ **Homologous** ___ series.

8.16.12 These are positional constitutional isomers.

8.16.13 These are skeletal constitutional isomers.

8.16.14

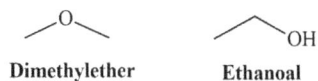

Dimethylether   **Ethanoal**

8.16.15

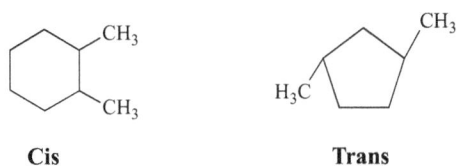

Cis   **Trans**

8.16.16

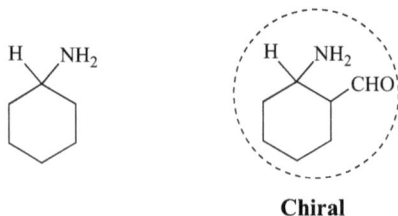

Chiral

8.16.17 It is achiral because it has two hydrogen atoms in its structure. For chiral amino acids there should be four different groups attached to central carbon atom.

8.16.18

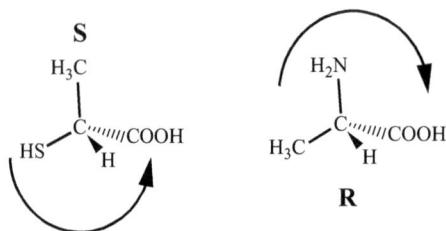

8.16.19

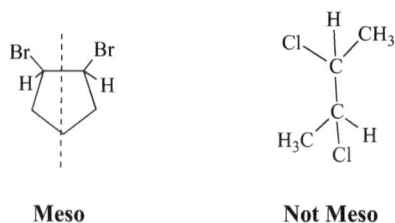

Meso   **Not Meso**

8.16.20 Yes

8.16.21

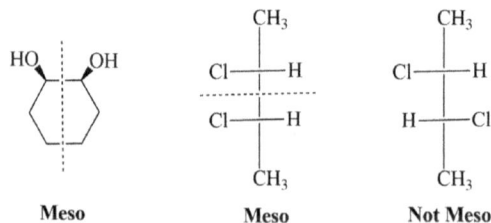

Meso   **Meso**   **Not Meso**

8.16.22 A racemic mixture is one that has equal amounts of left- and right-handed enantiomers of a chiral molecule. Simply it is a 50:50 mixture of enantiomers.

8.16.23 This molecule has a plane of symmetry (the horizontal plane going through the red broken line) and, therefore, is achiral; However, it has two chiral carbons and is consequentially a meso compound.

8.16.24 In phenols, the enol form, which is stabilized by the aromatic feature of the benzene ring, is the prominent tautomer.

8.16.25

Enol   Keto

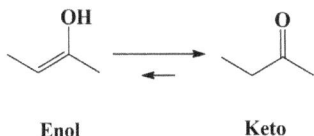

8.16.26 The ratio is a 50:50 or 1:1 equimolar mixture of two enantiomers.

8.16.27

C-4 Epimers

CHO	CHO
H——OH	H——OH
HO——H	HO——H
H——OH	HO——H
H——OH	H——OH
CH₂OH	CH₂OH

D-Glucose   **D-Galactose**

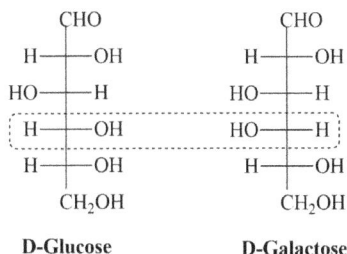

8.16.28 Achiral molecules are the ones that have their two mirror-image forms superimposable in three dimensions.

8.16.29 Yes, a racemic mixture can be created by combining an equal number of moles of two enantiomers.

8.16.30 Exo and endo are determined by the position of a substituent in the main ring relative to the shortest bridge. When the substituent is facing the bridge, the exo description is assigned to it. It is endo-configured while facing away from the bridge.

****

Chapter 9

9.6.1

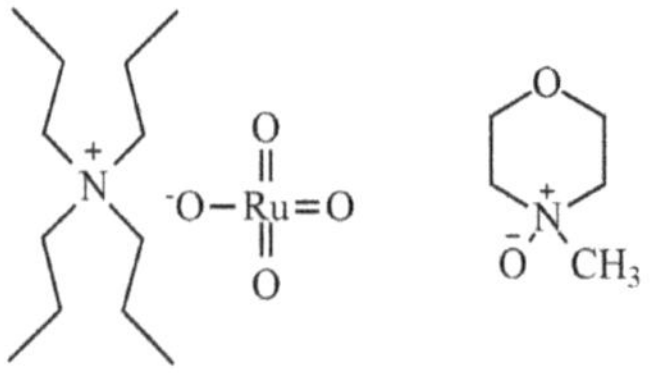

TPAP   NMO

9.6.2 Tetrabutylammonium iodide
9.6.3

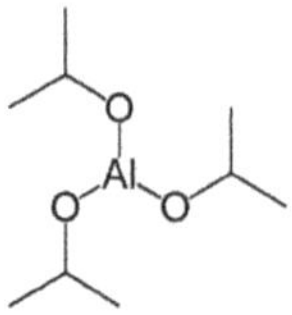

9.6.4 Pinacol is an organic compound that has two hydroxy groups
(–OH) on vicinal carbon atoms. The simplest example is 2,3-dimethyl-2,3-butanediol.

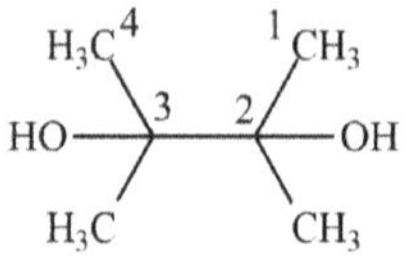

Pinacol

9.6.5 The pinacol-pinacolone rearrangement is an approach to convert a 1,2-diol to a carbonyl compound under acidic conditions

9.6.6

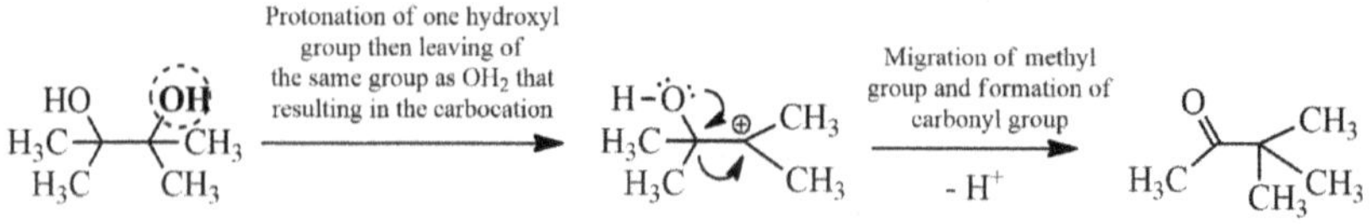

Protonation of one hydroxyl group then leaving of the same group as OH$_2$ that resulting in the carbocation

Migration of methyl group and formation of carbonyl group

- H$^+$

Pinacol   Pinacolone

9.6.7

Non-enolizable α-carbon
"No C-H bonds"
Only C-C bonds

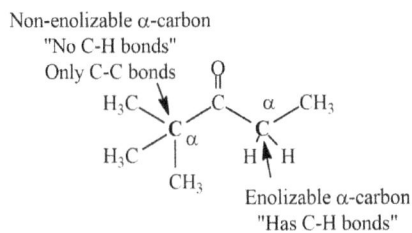

Enolizable α-carbon
"Has C-H bonds"

9.6.8

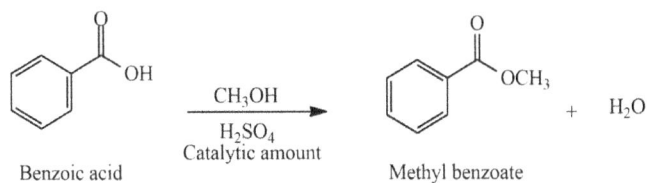

Benzoic acid $\xrightarrow[\substack{H_2SO_4 \\ \text{Catalytic amount}}]{CH_3OH}$ Methyl benzoate $+$ H_2O

9.6.9   Allyl and vinyl halides

9.6.10

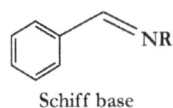

Schiff base

9.6.11   1,4-diazabicyclo[2.2.2]octane

9.6.12

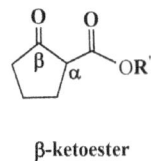

β-ketoester

9.6.13   Amalgamated zinc or zinc amalgam [Zn(Hg)] is the zinc that is surface-treated with mercury. It is commonly used for reduction reactions such as Clemmensen reduction of aldehydes and ketones to hydrocarbons

9.6.14

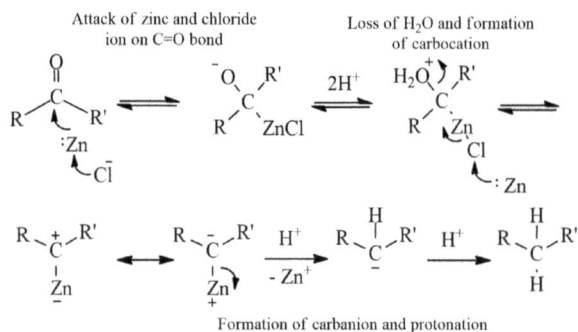

Attack of zinc and chloride ion on C=O bond

Loss of H$_2$O and formation of carbocation

Formation of carbanion and protonation

9.6.15 Esters containing both α-hydrogen and a carbonyl group

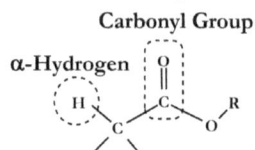

Carbonyl Group

α-Hydrogen

9.6.16

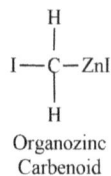

Organozinc
Carbenoid

9.6.17 Alkyne zipper reaction

9.6.18 An example of hydrogen donor is 1,4-cyclohexadiene (CHD) that is used in Bergman cyclization

1,4-Cyclohexadiene

9.6.19 Is the thermal isomerization of 1,5-dienes and related compounds in a reversible process

9.6.20 Tetramethylethylenediamine which is abbreviated as TMEDA or TEMED

9.6.21

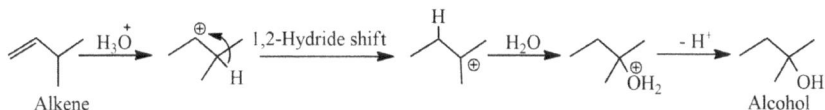

9.6.22 Boronic acid or boronic ester or in special cases with aryltrifluoroborane

9.6.23

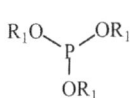

Trialkyl phosphite   Alkyl phosphonate
"Trivalent phosphorus "   "Pentavalent phosphorus"

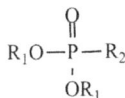

9.6.24 Palladium-catalyst

9.6.25 Ullmann reaction and Fittig reaction

9.6.26 Alkoxides and phenoxides

9.6.27 DEAD stands for diethyl azodicarboxylate and DIAD stands for diisopropyl azodicarboxylate.

9.6.28 Jones Reagent is aqueous acetone solution of CrO_3. It is used for the oxidation of primary alcohol to carboxylic acid and secondary alcohol to ketone.

9.6.29 The primary catalyst is tetrapropylammonium perruthenate (TPAP).

9.6.30 Glycol cleavage reaction can be achieved by several oxidizing agents including periodic acid (HIO_4), PIDA [$PhI\,(OAc)_2$], cerium (IV) salts, and silver(I) salts.

✱✱✱✱✱✱✱✱✱✱

Abbreviations

$Al_2(CO_3)_3$	Aluminum carbonate
$AlCl_3$	Aluminum chloride
NH_3	Ammonia
NH_4Cl	Ammonium chloride
BF_3	Boron trifluoride
Br_2	Bromine
Cl_2	Chlorine
$CrCl_2$	Chromium (II) chloride
CrO_3	Chromium trioxide
CrO_2Cl_2	Chromyl chloride
CuBr	Copper (I) bromide
CuCl	Copper (I) chloride or cuprous chloride
CHD	1,4-Cyclohexadiene
DAIB	(Diacetoxyiodo)benzene
N_2H_2	Diazene (nitrogen hydride)
DABCO	1,4-Diazabicyclo[2.2.2]octane
$(BH_3)_2$	Diborane
CH_2Cl_2 or DCM	Dichloromethane or methylene chloride
DEAD	Diethyl azodicarboxylate
DIBAL-H	Diisobutyl aluminum chloride
DIAD	Diisopropyl azodicarboxylate
DMF	Dimethylformamide
DMSO	Dimethylsulfoxide
DTT	Dithiothreitol
EG	Ethylene glycol
$FeCl_3$	Ferric chloride
HDPE	High density polyethylene
HBr	Hydrobromic acid or hydrogen bromide
HCl	Hydrochloric acid or hydrogen chloride
H_2O_2	Hydrogen peroxide
IUPAC	International Union of Pure and Applied Chemistry
$Pb\,(OAc)_4$	Lead tetraacetate
$LiAlH_4$	Lithium aluminum hydride
$MgCl_2$	Magnesium chloride
BME	β-Mercaptoethanol
$HgSO_4$	Mercury(II) sulfate
NMO	N-Methyl morpholine-N-oxide
$NiCl_2$	Nickel(II) chloride
HNO_3	Nitric acid
OsO_4	Osmium tetroxide
Pd	Palladium
$Pd\,(OAc)_2$	Palladium(II) acetate
Pd-C	Palladium on carbon
$Pd(PPh_3)_4$	Palladium-tetrakis(triphenylphosphine)
HIO_4	Periodic acid
PIDA [$PhI(OAc)_2$]	Phenyl iodine (III) diacetate
$POCl_3$	Phosphorus oxychloride

https://doi.org/10.1515/9783111382753-011

PBr3	Phosphorus tribromide
PCl_5	Phosphorus pentachloride
PE	Polyethylene
PMNA	Poly (methyl methacrylate)
PP	Polypropylene
PS	Polystyrene
PVAc	Polyvinyl acetate
PVC	Polyvinyl chloride
$K_2Cr_2O_7$	Potassium dichromate
KOH	Potassium hydroxide
KI	Potassium iodide
$KMnO_4$	Potassium permanganate
PCC	Pyridinium chlorochromate
PDC	Pyridinium dichromate
$NaBH_4$	Sodium borohydride
NaH	Sodium hydride
NaOH	Sodium hydroxide
$NaOCH_3$	Sodium methoxide
H_2SO_4	Sulfuric acid
Bu_4NBr or TBAB	Tetrabutylammonium bromide
TBAF or Bu_4NF	Tetrabutylammonium fluoride
HBF_4	Tetrafluoroboric acid
THF	Tetrahydrofuran
TMEDA/TEMED	Tetramethylethylenediamine
TPAP/TPAPR	Tetrapropylammonium perruthenate
$SOCl_2$	Thionyl chloride
3D	Three-dimension
$SnCl_2$	Tin (II) chloride or stannous chloride
Et_3N	Triethylamine
TASF	Tris(dimethylamino)sulfonium difluorotrimethylsilicate

Resources and Further Readings

Books

[1] Stoker, H. S. General, Organic, and Biological Chemistry, Brooks Cole, 2010, ISBN: 9780618606061.
[2] McMurry, J., Castellion, M. E., Ballantine, D. S., Hoeger, C. A., and Virginia, E. Fundamentals of General, Organic, and Biological Chemistry, Peterson, 2007, ISBN: 0136054501.
[3] Solomons, T. W. G., Fryhle, C. B., and Snyder, S. A. Organic Chemistry, John Wiley & Sons, 2016, ISBN: 9781118875766.
[4] Zweifel H. Plastics Additives Handbook, Hanser Publications, 2001, ISBN: 3446216545.
[5] Carraher, C. E. Seymour Carraher's Polymer Chemistry, CRC Press, Taylor & Francis Group, 2006, ISBN: 9781420051025.
[6] Wade, L. G. Organic Chemistry, Pearson, 2014, ISBN: 9781292021652.
[7] McMurry, J. Organic Chemistry, Brooks Cole, 2011, ISBN: 9780840054449.
[8] Morrison, R. T., and Boyed, R. N. Organic Chemistry, Prentice Hall, 1992, ISBN: 0136436692.
[9] Joule, J. A., and Mills, K. Heterocyclic Chemistry, John Wiley & Sons Ltd, 2010, ISBN: 9781405193658.
[10] Atkins, R. C., and Carey, F. A. Organic Chemistry "A Brief Course", McGraw-Hill, 2002, ISBN: 0072319445.
[11] Vollhardt, P., and Schore, N. Organic Chemistry "Structure and Function", Freeman and Company, 2014, ISBN: 139871464120275.
[12] Brown, W. H., Iverson, B. L., Anslyn, E. V., and Foote, C. S. Organic Chemistry, Wadsworth, Cengage Learning, 2014, ISBN: 139871285052816.
[13] Brown, W. H., and Poon, T. Organic Chemistry, John Wiley & Sons, Inc, 2016, ISBN: 9781118-875803.
[14] Solomons, T. W. G., Fryhle, C. B., and Snyder, S. A. Organic Chemistry, John Wiley & Sons, Inc, 2013, ISBN: 9781118147399.
[15] Elzagheid, M. Macromolecular Chemistry: Natural & Synthetic Polymers, Walter de Gruyter GmbH & Co KG, 2021, ISBN: 9783110762754.
[16] Greene, T. W., and Wuts, P. G. M. Protective Groups in Organic Chemistry, John Wiley & Sons, Inc, 1999, ISBN: 0471160199.
[17] Carey, F. A., and Sundberg, R. J. Advanced Organic Chemistry, Part A: Structure and Mechanisms, Springer, 2007, ISBN: 9780387448978.
[18] Carey, F. A., and Sundberg, R. J. Advanced Organic Chemistry, Part B: Reactions and Synthesis, Springer, 2007, ISBN: 9780387683508.

Internet Resources

— Branches of chemistry. Accessed on 10-10-2023. https://socratic.org/questions/what-are-the-branches-of-chemistry-and-their-definition.
— Meet the (Most Important) Functional Groups. Accessed on 11-10-2023. https://www.masterorganic chemistry.com/2010/10/06/functional-groups-organic-chemistry/
— Hybrid orbitals. Accessed on 11-10-2023. https://courses.lumenlearning.com/suny-potsdam-organic chemistry/chapter/2-2-hybrid-orbitals/
— The Nature of Chemical Bonds. Accessed on 12-10-2023. https://bio.libretexts.org/Bookshelves/Introduc tory_and_General_Biology/Map%3A_Raven_Biology_12th_Edition/02%3A_The_Nature_of_Molecules_ and_the_Properties_of_Water/2.03%3A_The_Nature_of_Chemical_Bonds

— Unsaturated Hydrocarbons. Accessed on 14-10-2023. https://www.angelo.edu/faculty/kboudrea/index_2353/Chapter_02_2SPP.pdf.

— Natural gas. Accessed on 14-10-2023. https://www.britannica.com/science/natural-gas.

— Petroleum. Accessed on 14-10-2023. https://education.nationalgeographic.org/resource/petroleum/

— LDPE and HDPE. Accessed on 14-10-2023. https://www.google.com/search?q=ldpe&sca_esv=573403102&rlz=1C1CHBF_enSA919SA919&sxsrf=AM9HkKmHMTQud4Qpz2GE__gaTPjTqIqQkw%3A1697269787099&ei=G0gqZfnVBayikdUPuaWJ0AU&ved=0ahUKEwi5z_3VhvWBAxUsUaQEHblSAloQ4dUDCBA&uact=5&oq=ldpe&gs_lp=Egxnd3Mtd2l6LXNlcnAiBGxkcGUyBxAAGIoFGEMyBxAAGIoFGEMyBxAAGIoFGEMyBxAAGIoFGEMyBRAAGIAEMgUQABiABDIFEAAYgAQyBRAAGIAEMgYQABgHGB4yBhAAGAcYHkj-EFAAWABwAHgBkAEAmAHAaAB5wGqAQMyLTG4AQPIAQD4AQHiAwQYACBBiAYB&sclient=gws-wiz-serp

— Crystalline structures in plastics. Accessed on 15-10-2023. http://www.dynamicscience.com.au/tester/solutions1/chemistry/crystallinestructures.html.

— amorphous, semi-crystalline, and crystalline polymer structure. Accessed on 15-10-2023. https://www.istockphoto.com/vector/vector-illustration-of-amorphous-semi-crystalline-and-crystalline-polymer-structure-gm1321630829-407807985.

— Fittig Reaction. Accessed on 16-10-2023. https://protonstalk.com/haloalkanes-and-haloarenes/fittig-reaction/

— Wurtz-Fittig Reaction Mechanism. Accessed on 16-10-2023. https://byjus.com/chemistry/wurtz-fittig-reaction-mechanism/.

— Alcohol-based Michaelis–Arbuzov reaction: an efficient and environmentally-benign method for C–P(O) bond formation. Accessed on 17-10-2023. https://pubs.rsc.org/en/content/articlelanding/2018/gc/c8gc00931g.

— Swarts Reaction. Accessed on 17-10-2023. https://unacademy.com/content/jee/study-material/chemistry/swarts-reaction/.

— Emmert Reaction. Accessed on 17-10-2023. https://www.drugfuture.com/OrganicNameReactions/ONR119.htm.

— Claisen Condensation. Accessed on 17-10-2023. https://chem.libretexts.org/Bookshelves/Organic_Chemistry/Organic_Chemistry_(Morsch_et_al.)/23%3A_Carbonyl_Condensation_Reactions/23.07%3A_The_Claisen_Condensation_Reaction.

— Baylis-Hillman Reaction. Accessed on 18-10-2023. https://www.sciencedirect.com/topics/chemistry/baylis-hillman-reaction.

— Forster Reaction. Accessed on 18-10-2023. https://www.drugfuture.com/OrganicNameReactions/ONR142.htm.

— Vilsmeier-Haack Reaction. Accessed on 18-10-2023. https://www.chemistrysteps.com/vilsmeier-haack-reaction/.

— Thiele Reaction. Accessed on 18-10-2023. https://www.drugfuture.com/OrganicNameReactions/onr393.htm.

— Ozonolysis Mechanism – Ozonolysis of Alkenes and Alkynes. Accessed on 18-10-2023. https://byjus.com/chemistry/ozonolysis-of-alkenes-alkynes-mechanism/.

— Difference Between Stereocenter and Chiral Center. Accessed on 19-10-2023. https://www.differencebetween.com/difference-between-stereocenter-and-chiral-center/.

— Stereoisomerism and Chirality. Accessed on 20-10-2023. https://slideplayer.com/slide/17951667/.

— Heck Reaction. Accessed on 22-10-2023. https://chem.libretexts.org/Bookshelves/Inorganic_Chemistry/Supplemental_Modules_and_Websites_(Inorganic_Chemistry)/Catalysis/Catalyst_Examples/Heck_Reaction.

— Hiyama Coupling. Accessed on 22-10-2023. https://www.organic-chemistry.org/namedreactions/hiyama-coupling.shtm.

— Chiral and Achiral Molecules. Accessed on 23-10-2023. https://socratic.org/organic-chemistry-1/r-and-s-configurations/chiral-and-achiral-molecules-1.
— Chirality. Accessed on 23-102023. https://www.chemistrylearner.com/chirality.html.
— Criegee Glycol Oxidation. Accessed on 23-10-2023. https://onlinelibrary.wiley.com/doi/10.1002/9780470638859.conrr168.
— Oppenauer Oxidation. Accessed on 23-10-2023. https://www.organic-chemistry.org/namedreactions/oppenauer-oxidation.shtm.
— Pinacol Pinacol-Pinacolone Pinacolone Rearrangement. Accessed on 23-10-2023. https://ddugu.ac.in/ePathshala_Attachments/STUDY298@266850.pdf.
— Alkyne Zipper Reaction. Accessed on 24-10-2023. https://synarchive.com/named-reactions/alkyne-zipper-reaction.
— Hydroboration–oxidation reaction. Accessed on 24-10-2023. https://en.wikipedia.org/wiki/Hydroboration%E2%80%93oxidation_reaction.
— Cope Rearrangement. Accessed on 25-10-2023. https://www.masterorganicchemistry.com/reaction-guide/cope-rearrangement/.
— Hay Coupling. Accessed on 25-10-2023. https://www.sciencedirect.com/topics/chemistry/hay-coupling.
— Kucherov Reaction. Accessed on 25-10-2023. https://www.drugfuture.com/Organic_Name_Reactions/topics/ONR_CD_XML/ONR229.htm.
— Oxymercuration Demercuration. Accessed on 25-10-2023. https://www.sciencedirect.com/topics/chemistry/oxymercuration-demercuration.
— Upjohn Dihydroxylation. Accessed on 25-10-2023. https://www.organic-chemistry.org/namedreactions/upjohn-dihydroxylation.shtm.
— What's a Racemic Mixture? Stereochemistry and Chirality. Accessed on 25-10-2023. https://www.masterorganicchemistry.com/2012/05/23/whats-a-racemic-mixture/.
— List of important Organic Compounds. Accessed on 01-11-2023. https://www.jagranjosh.com/general-knowledge/list-of-important-organic-compounds-1456306311-1.
— Common Reagents. Accessed on 06-11-2023. https://commonorganicchemistry.com/Sidebar/Common_Reagents.htm.
— List of organic compounds. Accessed on 17-11-2023. http://www.fullsense.com/Application/ChemicalOrganic/List_of_organic_compounds.htm#W.

Appendix A
List of Selected Important Organic Compounds

Organic compound	Use	Chemical structure
Acetaldehyde	Utilized in the manufacturing of antioxidants, rubber accelerators, and polymers.	
Acetamide	Used as a plasticizer and as an industrial solvent.	
Acetic acid	Mostly utilized in the manufacturing of synthetic fibers and textiles, polyvinyl acetate for wood glue, and cellulose acetate for photographic film.	
Acetic anhydride	Used to make synthetic or artificial silk from cellulose and to make medications like aspirin.	
Acetone	Low-grade acetone is a universal solvent that is widely used in academic laboratory settings as a glassware rinse agent to remove residue and solids prior to using a final wash. It is also an excellent precursor to methyl methacrylate.	
Acetylene	Used in metal welding, cutting, and heat treatment. It's also used to make synthetic rubber, neoprene, and polyethylene plastics.	
Aniline	In addition to being utilized as an organic solvent, it is also used in the condensation of methylenedianiline and related chemicals with formaldehyde.	
Benzaldehyde	Used in the production of dyes and also in the flavor and fragrance industries.	

https://doi.org/10.1515/9783111382753-013

(continued)

Organic compound	Use	Chemical structure
Benzene	Used to make synthetic fibers, rubber lubricants, and pigments.	
Benzoic acid	Used as antimicrobial preservative in food and beverages.	
Biotin	Helps in keeping skin, hair, eyes, liver, and nervous system healthy.	
Butane	Used as a solvent, in gasoline blending, and as a feedstock in the production of butadiene.	
Butanone	A common solvent utilized in cellulose acetate, nitrocellulose, and vinyl films.	
Carbon-tetrachloride	Used as a rubber industry solvent, a dry-cleaning industry cleansing agent, and a chemical and pharmaceuticals industry solvent.	
Chloroform	Utilized as an extraction solvent for fats, oils, greases, rubber, waxes, resins, and lacquers. It is also used in the production of artificial silk, gums, and adhesives.	
CPME "cyclopentyl methyl ether"	A good substitute for THF and dioxane. It can also be used for extraction, polymerization, crystallization, and surface coating.	
DMSO "dimethyl sulfoxide"	A common solvent that is miscible with water and a variety of organic solvents.	

(continued)

Organic compound	Use	Chemical structure
1,4-Dioxane	Employed as a solvent in a number of practical and scientific applications as well as a stabilizer for the transport of chlorinated hydrocarbons in aluminum containers.	
Diethyl ether	Used as a solvent in the synthesis of dyes and polymers.	
Ethyl acetate	Utilized as a process solvent in the pharmaceutical industry, an extraction solvent for a range of procedures, and a paint coating solvent.	
Ethyl alcohol	As important organic solvent used in artificial colors in perfumes and scent of fruits. It is also used as a form of fuel for motor vehicle.	
Formaldehyde	Used in producing fertilizers, paper, plywood, and certain resins. It is also used to preserve food and in household products.	
Formic acid	Used as a preservative and antibacterial agent as well as in the production of leather and rubber.	
Glycerol	Its industrial applications include nitroglycerin, ointment bases, solvents, food preservatives, plasticizers, and humectants.	
Hexane	Used as a multipurpose solvent and oil extractant. It is also utilized as a comonomer in polyethylene manufacturing.	
Ibuprofen	Used as nonsteroidal anti-inflammatory medication to treat inflammation, fever, and pain.	

(continued)

Organic compound	Use	Chemical structure
Isoprene	Used in the production of block polymers comprising butyl, isoprene, and styrene and also pressure-sensitive adhesives.	
Jenner's stain "methylene blue eosinate"	Used in microscopy for staining blood smears.	
Kanamycin	An antibiotic used to treat severe bacterial infections and tuberculosis.	
Lactic acid	Polymer precursor used in the production of the biodegradable polyester polylactide (PLA).	
Mecoprop "2-(2-methyl-4-chlorophenoxy) propionic acid"	Often employed as a broad herbicide.	
Methyl alcohol	Used as a solvent and in making plastics, polyesters, and as precursor for other chemicals such as formaldehyde and acetic acid.	
Naphthalene	Used as a raw material in the production of phthalic anhydride, which is used in the production of dyes, and plasticizers.	
Oxalic acid	Used to clean metals and objects of spots and rust.	

(continued)

Organic compound	Use	Chemical structure
Phenol	Used as precursor for plastics, and also versatile precursor to a large collection of drugs and many herbicides.	
Quinoline	Used as a solvent, and as a raw material for manufacture of dyes, and antiseptics.	
Pyridine	Used as a polar, basic, low-reactive solvent.	
Resorcinol	Use as antiseptic and disinfectant in topical pharmaceutical products. It is also used as a reagent in the carbohydrates qualitative tests.	
Retinol "vitamin A1"	Used as a dietary supplement.	
Ruthenium(III) acetylacetonate	An effective reusable catalyst that works well in clean circumstances to acetylate amines, alcohols, and phenols.	
Sildenafil	Used for the treatment of erectile dysfunction.	
Styrene	Used to make latex, synthetic rubber, and polystyrene resins.	

(continued)

Organic compound	Use	Chemical structure
Sudan III	A fat-soluble dye used for demonstrating triglycerides in frozen sections.	
THF	Used to dissolve polymers prior to employing gel permeation chromatography to determine their molecular mass. It is a typical solvent for Grignard reagents, and organometallic compounds like organolithium, and hydroboration processes.	
Toluene	Used as a precursor to benzene and xylenes, as a solvent for thinners, paints, lacquers, adhesives, and as an additive for gasoline.	
Urea	Used as fertilizer, and in making formaldehyde and urea plastic.	
Vanillin	Used as a general-purpose stain for visualizing spots on thin-layer chromatography plates. It is also used in the flavor industry.	
Warfarin	An oral anticoagulant commonly used to treat and prevent blood clots.	
Xylene	Used as a solvent and also common component of ink, rubber, and adhesives.	

(continued)

Organic compound	Use	Chemical structure
Yohimbine	Used to treat erectile dysfunction. It helps with fat loss among bodybuilders.	
Zingiberene	Used in traditional medicine for the treatment of inflammatory diseases, tumors, and bacterial infections.	

Appendix B
List of Selected Organic Reagents

Abbreviation	Full name	Uses
ADDP	1,1′-(Azodicarbonyl)dipiperidine	A common reagent in the Mitsunobu reaction.
AIBN	Azobisisobutyronitrile	A common reagent used to initiate radical reactions.
9-BBN	9-Borabicyclo[3.3.1]nonane	Commonly utilized in hydroboration-oxidation reaction.
BOP	Benzotriazol-1-yloxytris (dimethylamino)phosphonium hexafluorophosphate	Widely employed in peptide synthesis
CAN	Ceric ammonium nitrate	A strong oxidizing agent and oxidant in quantitative analysis.
CMPI	2-Chloro-1-methylpyridinium iodide	Widely used as an effective and adaptable organic condensation agent.
DABCO	1,4-diazabicyclo[2.2.2]octane	A catalyst, an organic base, and a complexing ligand.
DAST	Diethylaminosulfur trifluoride	A fluorinating agent that is widely used in organic chemistry.
DCC	Dicyclohexylcarbodiimide	A coupling reagent used in the formation of amide bonds.
DEAD	Diethyl azodicarboxylate	Reagent that is commonly used in Mitsunobu reactions.
DIAD	Diisopropyl azodicarboxylate	Employed in the Mitsunobu reaction as an oxidant of triphenylphosphine-to-triphenylphosphine oxide.
DIBAL-H	Diisobutylaluminum hydride	An organoaluminum reagent that is commonly used to reduce esters or nitriles to aldehydes.
DMAP	4-(Dimethylamino)pyridine	A nucleophilic esterification and hydrosilylation catalyst.
Dppb	[1,4-Bis(diphenylphosphino)butane]	A ligand that is commonly utilized in palladium chemistry as well as other transition metal chemistry such as nickel or rhodium.

https://doi.org/10.1515/9783111382753-014

(continued)

Abbreviation	Full name	Uses
Dppe	1,2-Bis(diphenylphosphino)ethane	A bidentate ligand that is commonly utilized in coordination chemistry.
HATU	Hexafluorophosphate azabenzotriazole tetramethyl uronium	Utilized to create an active ester from a carboxylic acid in peptide coupling chemistry.
HMPA	Hexamethylphosphoramide	Used as a solvent or additive on occasionally.
KHMDS	Potassium bis(trimethylsilyl)amide	Strong, non-nucleophilic base used for deprotonations.
LAH	Lithium aluminum hydride	Utilized in chemical synthesis as a reducing agent, particularly for the reduction of esters, carboxylic acids, and amides.
MSA	Methanesulfonic acid	Useful as a catalyst in organic reactions, particularly polymerizations, esterifications, and transesterifications.
MTB	Thioanisole or (methylthio)benzene	Utilized as a precursor in the production of dyes, medicines, and agrochemicals
NBS	(N-bromosuccinimide)	A common reagent in radical substitution and electrophilic addition processes.
NIS	N-iodosuccinimide	Iodination reagent.
NMO	N-methylmorpholine N-oxide	Co-oxidant in the conversion of alkenes to diols using OsO_4.
PCC	Pyridinium chlorochromate	Used for the oxidation of alcohols to ketones.
PDC	Pyridinium dichromate	A strong oxidizing agent is used to convert primary alcohols to aldehydes and secondary alcohols to ketones.
PPTS	Pyridinium p-toluenesulfonate	Acidic catalyst for THP protection.
SMEAH	Sodium bis(2-methoxyethoxy)-aluminum hydride	A versatile agent for reducing hydrides. It easily transforms epoxides, anhydrides, acyl halides, ketones, carboxylic acids, and esters into their corresponding alcohols.
STAB	Sodium triacetoxyborohydride	A mild reducing agent is utilized in reductive aminations.

(continued)

Abbreviation	Full name	Uses
TBAF	Tetrabutylammonium fluoride	Utilized to get rid of protective groups for silyl ether. It also serves as a moderate base and a phase transfer catalyst.
TFAA	Trifluoroacetic anhydride	Used to convert amides to nitriles.
TMAF	Tetramethylammonium fluoride	Used to convert aryl phenols to aryl fluorides when combined with sulfuryl fluoride.
TMEDA	Tetramethylethylenediamine	Used as a polymerization accelerator in gel electrophoresis as well as a solvent and oxidizing reagent.
TPAP	Tetrapropylammonium perruthenate	Used to convert primary alcohols to carboxylic acids.
UHP	Urea hydrogen peroxide	Used to convert nitriles to primary amides.

Index

1,2-ethanediol 45
2-amino-5-guanidinopentanoic acid 76, 99
3-D 94

absolute configuration 99
achiral molecule 96
acid anhydrides 80
acid chlorides 42, 81
acrylonitrile 23
acyl group 78
addition and condensation 87
addition polymerization 88
adenine 54
adipic acid 77
alcohols 43
aldehydes 62
aliphatic 14
alkane 20
alkenes 20
alkoxymercuration 50
alky group 17
alkyl 17
alkyl and arylamines 40
alkyl azides 41
alkylamines 40
alkyne zipper reaction 111
alkynes 15, 20
alphabetical order 30
amides 81
amino carboxylic acids 75
aniline 42
anthracene 31
arachidonic acid 75
Arbuzov reaction 117
aromatic 14
aromatic compounds 29
arylamines 40
aspartic acid 76
asymmetrical 40
atactic 89
aza 54

Baylis-Hillman reaction 111
benzene 28
benzene derivative 36
Bergman cyclization 111

BME 47
branched 85
branches of organic chemistry 6
bromine water 35
Brown hydroboration 111
Buchwald-Hartwig reaction 117

Cahn-Ingold-Prelog sequence rules 99
Cannizzaro Reaction 68
capric acid 73
caprylic acid 73
carboxylic acids 71
carcinogenic 32
chiral center 98
cis isomers 104
Claisen condensation 111
Clemmensen reduction 111
combustion 20
commodity plastics 89
condensation polymerization 88
Condensed 1
constitutional isomers 94
Cope rearrangement 111
copolymer 90
coronene 31
covalent bonds 8
crown ethers 48
crude oil 22
crystalline 87
crystallites 87
cumene 35
cycloalkanes 17
cytosine 54

demercuration 50
dialkyl sulfides 51
diastereoisomers 94
dicarboxylic acids 76
Dieckmann condensation 111
Diels-Alder 20
dienes 15, 20
dihalides 21
dipole-dipole interactions 62
direct halogenation of alkanes 34
dithiothreitol 47
dithiothreitolss 47

https://doi.org/10.1515/9783111382753-015

double 15
Dow process 34
DTT 47

electrophilic aromatic substitution 32
electrovalent (ionic) 8
Emmert reaction 111
enanthic acid 73
enantiomers 94
epimerization 103
epimers 103
esters 80
ethene 23
ethers 48
ethylene glycol 45
exo or endo 105

fatty acids 74
Fittig reaction 117
formic acid 72
Forster reaction 111
free radical halogenation 34
Friedel-Crafts 58
Friedel-Crafts alkylation 34
functional groups 1
furan 57
furfural 54
fused aromatic rings 31

Gattermann-Koch reaction 65
Glaser-Eglinton-Hay coupling 111
glutaric acid 77
Grignard reagent 64
guanine 54

haloalkanes 18
halobenzenes 33
halogen 18
halogenation 19–20
Heck reaction 117
heterocyclic 51
heterocyclic compound 54
hexadecenoic acid 75
hexanoic acid 73
Hiyama Coupling reaction 117
hybridization 6
hydration 21

hydrobromic acid 50
hydrocarbons 14–15
hydrohalogenation 21
hydroiodic acid 50
hydroxybenzene 33
hydroxyl group 15

indole 53
initiator 88
internal alkynes 20
isopropyl benzene 35
isoquinoline 53
isotactic 89
IUPAC 15
IUPAC naming system 19

Kekulé 1
ketones 62
Kucherov reaction 111

lactones 78
linear 85
lithium aluminum hydride 41

macromolecules 85
malonic acid 77
Markovnikov 19
mercaptans 45
meso compounds 102
meta 29
molecular chirality 96

$NaBH_4$ 50
naphthalene 31
natural gas 21
nickel 20
nitration 34
nitriles 82
Nozaki-Hiyama-Kishi 111
nucleic acids 53
– nucleic acids bases 53
nylon 6,6 88

octadeca-9,12-dienoic acid 75
octanoic acid 73
orbitals 6
Organic chemistry 1

ortho 29
oxa 54
oxalic acid 77
oxymercuration 21
ozonolysis 64, 111

palladium 20
palmitoleic acid 75
para 29
pelargonic acid 73
petroleum 22
phenanthrene 31
phenol 34
phenyl 29
phthalic acid 77
pi-bond 14
platinum 20
polyhydroxy alcohols 45
polymer architecture 85
polymer morphology 86
polymerization 87
polymers 85
polypropylene 85
primary 18
propagating 89
propene 23
pyrene 31
pyridine 57
pyridinium dichromate 64
pyrolysis 20
pyrrole 57

quinoline 53

racemic mixture 99

saponification 80
saturated 15, 74
secondary 18
secondary amines 42
serine 76
Simmons-Smith reaction 111
Skeletal 1
SnCl$_2$ 42
sp hybrid 8
sp orbitals 8
sp^2 hybrids 7
sp^3 hybrids 6

sp^3-hybridized 43
stearic acid 75
stereocenter 98
stereochemistry 94
stereoisomers 94
stereoscriptors 102
structural isomers 94
styrene 23
subfields 5
substituents 29
sugar alcohols 45
sulfide 51
sulfides 51
sulfonation 34
sulfur analogs 45
Suzuki-Miyaura reaction 117
Swarts reaction 117
Swern oxidation 64
symbol (R) 17
symmetrical 40
syndiotactic 89

tautomerism 106
terminal alkynes 20
tertiary 18
tertiary amines 40
tetrafluoroethene 23
tetrahydrofuran 57
THF 57
thia 54
Thiele reaction 111
thiol 45
thiolate anions 51
thiols 46
thionyl chloride 81
thiophene 57
three-dimensional 94
thymine 54
tin (II) chloride 42
trans isomers 104
triple 15

Ullmann reaction 117
unsaturated 74
unsymmetrical 48
Upjohn dihydroxylation 111
uracil 54

valine 76
Vilsmeier-Haack reaction 111
vinyl chloride 23

Williamson ether synthesis 50
Williamson reaction 117

Williamson synthesis 50
Wolff-Kishner reduction 111
Wurtz-Fittig reaction 117

β-mercaptoethanol 47

www.ingramcontent.com/pod-product-compliance
Lightning Source LLC
Chambersburg PA
CBHW081525220326
41598CB00036B/6336